Reviews of

116 Physiology Biochemistry and Pharmacology

Editors

M. P. Blaustein, Baltimore · O. Creutzfeldt, Göttingen
H. Grunicke, Innsbruck · E. Habermann, Gießen
H. Neurath, Seattle · S. Numa, Kyoto
D. Pette, Konstanz · B. Sakmann, Heidelberg
M. Schweiger, Innsbruck · U. Trendelenburg, Würzburg
K. J. Ullrich, Frankfurt/M · E. M. Wright, Los Angeles

With 24 Figures and 5 Tables

Springer-Verlag Berlin Heidelberg GmbH

ISBN 978-3-662-30985-8   ISBN 978-3-540-47169-1 (eBook)
DOI 10.1007/978-3-540-47169-1

Library of Congress-Catalog-Card Number 74-3674

© Springer-Verlag Berlin Heidelberg 1990
Originally published by Springer-Verlag Berlin Heidelberg New York in 1990
Softcover reprint of the hardcover 1st edition 1990

Typesetting: K+V Fotosatz GmbH, Beerfelden
2127/3130-543210 – Printed on acid-free paper

Contents

Rev. Physiol. Biochem. Pharmacol., Vol. 116
© Springer-Verlag 1990

Cellular and Molecular Diversities of Mammalian Skeletal Muscle Fibers

DIRK PETTE[1] and ROBERT S. STARON[2]

Contents

[1] Fakultät für Biologie, Universität Konstanz, D-7750 Konstanz, FRG
[2] Department of Zoological and Biomedical Sciences, College of Osteopathic Medicine, Athens, Ohio 45701, USA

1 Introduction

A unique characteristic of skeletal muscle is its diversity. This diversity is created by its design, i. e., its fiber composition and the heterogeneity of the individual fibers. Indeed, we now know that no one skeletal muscle within an animal is identical with a second. Even homologous muscles exhibit differences in fiber composition between species and strains. This heterogeneity of muscle tissue reflects its high degree of functional specialization and is the basis of its functional plasticity. The full extent of the diversity exhibited by muscle fibers is only now being realized. Current and future research is important for improving our understanding of the molecular basis of phenotypic expression in skeletal muscle fibers during development, during functional specialization, and under both normal and pathological conditions.

Previous reviews have dealt with various aspects of this diversity (Buchthal and Schmalbruch 1980; Burke 1981; Obinata et al. 1981; Gröschel-Stewart and Drenckhahn 1982; Eisenberg 1983; Saltin and Gollnick 1983; Bandman 1985b; Schmalbruch 1985; Gauthier 1986; Ohtsuki et al. 1986; Swynghedauw 1986; Syrovy 1987; Ogata 1988). The aim of the present review is to summarize and integrate our current knowledge of the qualitative and quantitative differences between adult skeletal muscle fibers at the levels of cellular and molecular organization. The vast amount of knowledge being accumulated in the separate disciplines of histochemistry, electron microscopy, biochemistry, biomechanics, and molecular biology needs to be integrated to achieve a more complete understanding. This review, by no means all encompassing, focuses mainly on adult mammalian skeletal muscle. Some reference is also made to avian muscle.

2 Muscle Fiber Types

2.1 Historical Background of Fiber Classification

Skeletal muscle fiber diversity was realized as early as 1873 when Ranvier distinguished "white" and "red" muscles. Further descriptions of the differences between individual muscle fibers ("clear fibers" and "opaque fibers") emerged from the histological studies of Grützner (1883) and Knoll (1891), who observed variations in fiber size and translucency. Years later, structural differences delineating two major types of fibers, *Fibrillenstruktur* fibers and *Felderstruktur* fibers, were observed by Krüger and coworkers (for a review see Krüger 1952). Finally, a breakthrough in the delineation of fiber types resulted from the combination of histological and physiological methods (Denny-Brown 1929; Henneman and Olson 1965), and from enzyme histochemical studies (Padykula and Herman 1955; Ogata 1958a–c; Dubowitz and Pearse 1960).

2.2 Histochemical Fiber Typing

At the moment histochemical fiber classification schemes depend upon subjective evaluation and are limited by the restricted resolution of the methods applied. Understandably, only a small number of major fiber types can be discriminated by qualitative histochemical methods. Nevertheless, histochemical fiber typing has proven useful in numerous studies concerned with structure – function relationships in normal and pathological muscles.

Presently, two different histochemical approaches allow for the separation of fiber types. One method is based upon myofibrillar actomyosin adenosine triphosphatase (ATPase) activity and the other upon reference enzymes of anaerobic and aerobic energy metabolism. Initial distinction of two myofibrillar actomyosin ATPase (mATPase)-based fiber types, type I and type II (Engel 1962), was made possible by the use of a histochemical assay for ATPase activity (Padykula and Herman 1955). Shortly thereafter, Bárány et al. (1965) demonstrated that fast and slow muscles contain myosins differing in their specific actin-activated and Ca^{2+}-dependent ATPase activities and that this correlates with differences in speed of contraction (Bárány 1967). It became clear that the histochemical differences in the activity of mATPase between type I and type II fibers correspond to differences in contractile properties (Close 1967, 1972; Barnard et al. 1971; Burke et al. 1971; Edgerton and Simpson 1971). Subsequently, differences in proteolytic fragments (Nakamura et al. 1971; Jean et al. 1973) and antigenicity (Arndt and Pepe 1975) suggested the existence of myosin heavy chain isoforms in white and red muscles.

Different types of muscle fibers have also been distinguished on the basis of histochemical reactions for enzymes of aerobic oxidative metabolism, e. g., succinate dehydrogenase, cytochrome oxidase, and nicotinamide adenine dinucleotide (NADH) tetrazolium reductase (Ogata 1958a–c). Three major fiber types derived from differences in these enzyme activities (Ogata and Mori 1964; Padykula and Gauthier 1967; Gauthier 1969, 1974) basically reflect differences in mitochondrial content (high, intermediate, low) and therefore primarily relate to differences in aerobic oxidative potentials (Hoppeler 1986, 1990; Hoppeler et al. 1987). Various fiber types can also be delineated histochemically by the inverse relationship between the activity of glycogen phosphorylase or mitochondrial glycerolphosphate dehydrogenase and the activity of aerobic oxidative metabolic enzymes (Dubowitz and Pearse 1960).

2.2.1 A Metabolic Enzyme-Based Scheme of Fiber Classification

Although enzyme histochemical methods made it possible to distinguish a limited number of fiber types, it remained unclear what relationships, if any, existed between the mATPase-based and metabolic enzyme-based classifications of fiber types. Such relationships became evident from the studies of Barnard et al. (1971) and Peter et al. (1972). They investigated the distribution of enzymes related to either aerobic oxidative or anaerobic glycolytic metabolism in fibers with low (type I) or high (type II) mATPase activity. NADH tetrazolium reductase and succinate dehydrogenase were chosen as reference enzymes for aerobic oxidative metabolism, and glycerolphosphate oxidase (α-glycerolphosphate dehydrogenase) was chosen as a reference enzyme for the anaerobic glycolytic pathway. Although glycerolphosphate oxidase is a mitochondrial, structure-bound flavoprotein, it is related to the glycolytic pathway via its function in the glycerolphosphate shuttle; it is present in amounts directly proportional to glycolytic enzyme activities in mammalian muscle (Pette and Bücher 1963; Pette 1966). This combination of metabolic enzyme-based and mATPase-based histochemical study resulted in the isolation of three major fiber types in muscles of guinea pig and rabbit. These fiber types were named slow-twitch oxidative (SO), fast-twitch oxidative glycolytic (FOG), and fast-twitch glycolytic (FG) (Barnard et al. 1971; Peter et al. 1972).

A method of depleting glycogen in single motor units made it possible to establish a relationship between metabolic properties and fatigability (Edström and Kugelberg 1968; Kugelberg and Edström 1968). It was shown that FG fibers represent fast-fatigable (FF) motor units, FOG fibers represent fast, fatigue-resistant (FR) motor units, and SO fibers represent slow, fatigue-resistant (S) motor units (Burke et al. 1971, 1973, 1974; Burke 1981). Thus, fatigue-resistant muscle fibers are rich in enzyme activities of aerobic substrate oxidation, regardless of whether they are fast or slow. In other words, red muscle fibers may either be fast or slow (Edgerton and Simpson 1969; Schmalbruch 1971; Peter et al. 1972).

2.2.2 Myofibrillar Actomyosin ATPase-Based Schemes
of Fiber Classification

The observation that fast and slow myosins have different alkaline and acid stabilities (Sréter et al. 1966; Seidel 1967) formed the basis for the elabora-

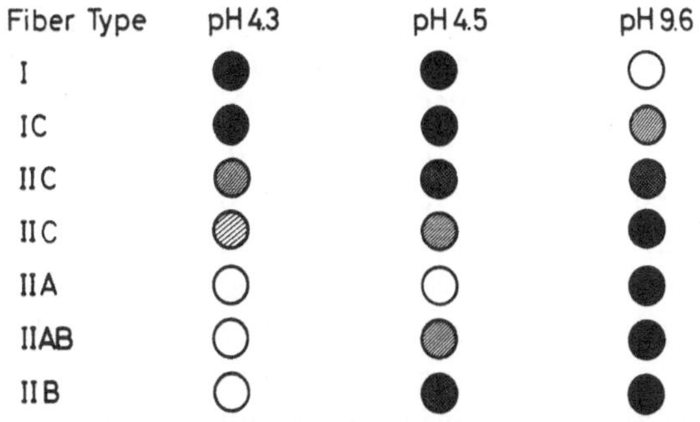

Fig. 1. Schematic illustration of histochemical fiber typing based upon mATPase staining intensity following preincubation at various pH values. (From Staron and Pette 1986)

tion of more refined methods for mATPase-based fiber type delineation. Histochemically, fast fibers display high mATPase activity under alkaline conditions and low activity under acid conditions (alkali-stable, acid-labile), whereas slow fibers exhibit the inverse (alkali-labile, acid-stable; Guth and Samaha 1969).

More detailed study of the acid stability of the mATPase activity indicated that the distinction into only two mATPase-based fiber types was an oversimplification and led to the delineation of fast fiber subtypes, i.e., types IIA, IIB, and IIC (Fig. 1; Brooke and Kaiser 1970a). Fiber type IIA stains deeply after alkaline preincubation (pH 10.4) and lightly after preincubations at pH 4.6 and pH 4.3. Fiber type IIB stains deeply after alkaline preincubation (pH 10.4), with intermediate intensity following preincubation at pH 4.6, and lightly after preincubation at pH 4.3. Type IIC fibers are stable to various degrees throughout the 4.3–10.4 pH range, which results in more or less intermediate reactions after preincubations at the three pH values.

Differences within the fast fiber population had also been noticed by Stein and Padykula (1962), based on the existence of different formaldehyde sensitivities of mATPase activity. Subsequently, Guth and Samaha (1969) and Samaha et al. (1970) used a combination of the different pH and formaldehyde sensitivities of mATPase activity to distinguish between slow fibers, termed β fibers, and two fast fiber types, termed α and $\alpha\beta$.

The two different mATPase-based classification schemes of Brooke and Kaiser (1970a,b) and of Guth and Samaha (1969) raise the question of compatibility. Because both systems distinguish two major subgroups of fast fibers (types IIA and IIB and types $\alpha\beta$ and α), it might be suggested that the two nomencleatures are interchangeable, i.e., IIA $= \alpha\beta$ and IIB $= \alpha$. In fact, many studies have used the method of Guth and Samaha (1969), but applied the nomenclature of Brooke and Kaiser (1970a). However, it must be taken into account that the delineation of the two groups of fast subtypes was accomplished by different methods, i.e., mATPase sensitivity towards pH (Brooke and Kaiser 1970a) or formaldehyde (Guth and Samaha 1969). Indeed, these classification systems do not appear to be entirely compatible. Green et al. (1982) showed complete correspondence between type I fibers and β fibers, but significant variations between the fast fiber subtypes IIA and IIB on the one hand and $\alpha\beta$ and α on the other hand. Furthermore, these differences were found to vary between species. Therefore, the two terminologies should not be interchanged, but should be used only with the methods from which they are derived (Green et al. 1982). This incompatibility may, perhaps, also be due to the existence of additional fast fiber subpopulations expressed in different muscles and different species.

Additional procedures for classification of fibers based on mATPase activity (Meijer 1970; Khan et al. 1972; Schiaffino and Pierobon-Bormioli 1973; Müntener 1979; Mabuchi and Sréter 1980; Round et al. 1980; Doriguzzi et al. 1983; Gollnick et al. 1983; Staron et al. 1983; Gollnick and Matoba 1984; Matoba and Gollnick 1984; Eddinger et al. 1985) represent modifications of the original procedure of Padykula and Herman (1955). These modifications have not only confirmed the existence of major fast fiber subtypes but have also identified more than one slow fiber type (Askanas and Engel 1975; Karpati et al. 1975; Khan et al. 1974; Khan 1978; Gollnick et al. 1983; Smith et al. 1987). The existence of additional fast fiber subtypes has also become evident (Andersen and Henriksson 1977; Jansson et al. 1978; Ingjer 1979; Green et al. 1979; Gollnick et al. 1983; Staron et al. 1983; Hughes 1986; Staron and Pette 1986; Fig. 1) and

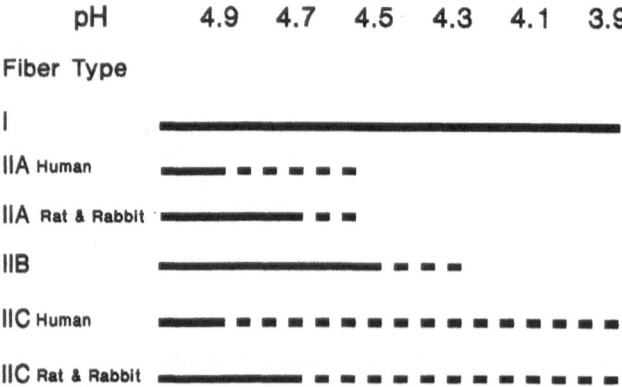

Fig. 2. Species differences in histochemically assessed myofibrillar ATPase activity after preincubation at various pH values. *Solid lines* indicate no inhibition. *Broken lines*, partial inhibition. Lines terminate at pH values at which inhibition is almost total. (According to Brooke and Kaiser)

has been supported by microphotometric evaluation of the histochemical reaction for mATPase activity (Vaage et al. 1980; Spurway 1981; van der Laarse et al. 1984; Spurway and Rowlerson 1989). Moreover, fiber types defined in one species may exhibit different histochemical properties in other species, e. g., the pH sensitivity of the mATPase activity (Fig. 2). Therefore, it may be necessary to appropriately modify the methods used for their histochemical distinction (Brooke and Kaiser 1970b; Dubowitz and Brooke 1973; Khan et al. 1974; Müntener 1979).

2.3 The Compatibility of Different Histochemical Classification Schemes

How mATPase-based classification systems relate to metabolic-based systems is an important question that needs to be addressed. The casual mixing of these different schemes has created confusion in the literature and among scientists not familiar with the limitations of each classification scheme. Additionally, it must be realized that fiber type classification schemes have been derived using relatively few muscles from a small number of species. Consequently, a given classification scheme derived using adult muscle may not even apply to different muscles within the same animal and may not be relevant to developing muscle, conditions of degeneration/regeneration, or altered phenotype expression (Guth and Samaha 1972; Dubowitz and Brooke 1973).

In view of the popularity of the methods of Brooke and Kaiser (1970a, b) and Barnard et al. (1971), the following discussion focuses on a comparison between these two classification systems. The assumption that metabolic properties should correspond to properties related to different mATPase activities may not be justified. Several histochemical studies employing qualitative procedures have compared fibers typed using both the Brooke and Kaiser (1970a) and the Barnard et al. (1971) methods (Sjögaard et al. 1978; Nemeth et al. 1979; Nemeth and Pette 1980, 1981a) and have indicated their incompatibility. These findings have been confirmed and extended by the use of microphotometry. Quantifying the histochemically assessed succinate dehydrogenase activity (a marker of aerobic oxidative metabolism) in type I, type IIA, and type IIB fibers reveals an entire spectrum of metabolic activity (Nemeth and Pette 1981b; Pette and Tyler 1983; Reichmann and Pette 1982, 1984).

Depending upon the species, varying amounts of overlap exist between the three fiber types (types I, IIA, IIB) with regard to their aerobic oxidative capacities. Based upon our present understanding, it appears incorrect to separate IIB and IIA fibers according to their different metabolic properties or to make predictions on their metabolic properties (Pette 1985). In addition, metabolic enzyme activity levels of the different fiber types depend upon the "training state" of the muscle (Baldwin et al. 1972; Gollnick et al. 1972, 1973; Winder et al. 1974; Green et al. 1983; Saltin and Gollnick 1983; Holloszy and Coyle 1984; Howald et al. 1985) and may rapidly change in response to altered functional demands because of the short half-lives of these enzymes (Dölken and Pette 1974; Illg and Pette 1979). Metabolic plasticity has been extensively studied at the level of enzyme activities involved in anaerobic and aerobic pathways of energy metabolism by the use of chronic nerve stimulation (Pette et al. 1983; Pette and Tyler 1973; Pette 1984; Reichmann et al. 1985; Henriksson et al. 1986; Seedorf et al. 1986; Simoneau and Pette 1988; Hood et al. 1989; Hood and Pette 1989).

Although the mean values of succinate dehydrogenase activity can be used to separate IIB and IIA fibers in rabbit, rat, guinea pig, and cat (IIA fibers being "oxidative"), some type IIB fibers are as oxidative as IIA fibers, and some IIA fibers exhibit succinate dehydrogenase activities as low as those exhibited by the majority of IIB fibers (Fig. 3; Nemeth and Pette 1981 b; Reichmann and Pette 1982; Pette and Tyler 1983). Pronounced overlap between type IIA and type IIB fibers with regard to succinate dehydrogenase activity is found in human (Reichmann and Pette 1982) and in horse (White and Snow 1985) muscles. Morphometric analyses of the mitochondrial distribution in human fiber types IIA and IIB also show pronounced overlaps between these two fiber populations (Sjöström et al. 1982 a; Staron et al. 1983; Hoppeler 1986). Moreover, the majority of IIB fibers in mouse hindlimb muscles display a higher aerobic oxidative capacity compared with the majority of IIA fibers (Sher and Cardasis 1976; Reichmann and Pette 1982, 1984; van der Laarse et al. 1984).

It is possible that the incompatibility of the mATPase-based and metabolic enzyme-based classification schemes may be partially explained by the existence of additional fast fiber subtypes. An additional fast fiber subtype (type 2X) has recently been delineated immunohistochemically (Schiaffino et al. 1985, 1986 c, 1988 a, 1989, 1990; Gorza 1990) in specific muscles of small mammals. The mATPase activity of this new fast fiber type exhibits an acid stability similar to that of type IIB fibers according to the procedure of Brooke and Kaiser (1970 a), but is distinguishable from type IIB after alkaline preincubation according to the procedure of Guth and Samaha (1969). In addition, this fiber type appears to be histochemically characterized by an aerobic oxidative capacity intermediate between that of the type IIB and that of type IIA fibers (Termin et al. 1989 a; Gorza 1990; Parry and Zardini 1990; Schiaffino et al. 1985, 1986 c, 1988 a, 1989, 1990). Thus, some type 2X fibers may have been erroneously classified as

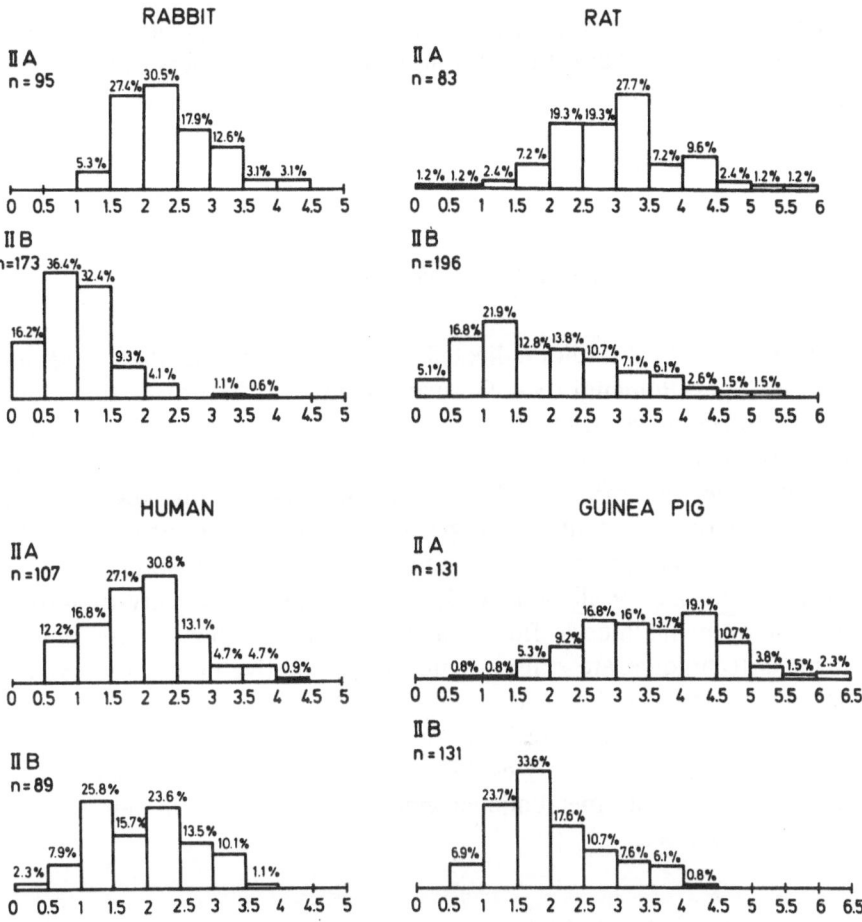

Fig. 3. Relative succinate dehydrogenase activities microphotometrically determined in type IIA and type IIB fibers of rabbit, guinea pig, rat, and human muscles. (Modified from Reichmann and Pette 1982)

type IIB in previous studies (e.g., Sher and Cardasis 1976; Reichmann and Pette 1982, 1984). One may also speculate that fast-twitch muscles of certain species (e.g., dog) which contain histochemically atypical IIB fibers (Snow et al. 1982) are made up by type 2X fibers.

An incompatibility between mATPase-based and metabolic enzyme-based classification schemes also appears to exist for (slow-twitch) type I fibers. In some mammals, e.g., human and cat, the type I fibers are the most oxidative and exhibit the highest activity levels of enzymes representing aerobic oxidative metabolic pathways, whereas in other mammals, e.g., rat, guinea pig, and horse, the type I fibers are intermediate in their oxidative potential (Barnard et al. 1971; Peter et al. 1972; Prince et al. 1976;

Reichmann and Pette 1982; Sjöström et al. 1982a; Hoppeler et al. 1983; Essén-Gustavsson 1986). In addition, metabolic differences exist between slow fibers in different muscles of the same animal (Spamer and Pette 1977). Therefore, the classification of all type I fibers as slow oxidative fibers may not be correct.

3 Biochemical Analyses on Single Muscle Fibers

Early histochemical studies indicated the existence of more than three major fiber types (Romanul 1964; Guth and Yellin 1971). However, the resolution of conventional histochemical methods is not sufficient to precisely define biochemical fiber type diversities. A better approach to elucidate fiber heterogeneity is offered by immunohistochemical techniques, although they depend on the availability of specific antibodies raised against specific antigens. This may present a problem considering that the source for antigen isolation may be ill defined due to the cellular diversity of muscle (nonhomogeneous muscle fiber population). In addition, immunohistochemical techniques are still evaluated by qualitative means. One of the more promising techniques for analyzing the molecular basis of fiber diversity is single fiber dissection for quantitative microbiochemical or microimmunochemical analyses. These microanalytical techniques provide valuable insights into metabolic enzyme and myofibrillar protein profiles of individual fibers.

3.1 Enzymes of Energy Metabolism

Comparative measurements of the activities of enzymes involved in energy metabolism have been used as tools to segregate various muscle types (Vogell et al. 1959; Pette and Bücher 1963; Dawson and Romanul 1964; Opie and Newsholme 1967; Bass et al. 1969; Burleigh and Schimke 1969; Crabtree and Newsholme 1972; Peter et al. 1972; Staudte and Pette 1972; Pette and Dölken 1975). Although enzyme activities determined in vitro cannot be set equal to metabolic flux rates in vivo, they may be used as relative measures of specific metabolic capacities (Pette and Bücher 1963; Pette 1966; Bass et al. 1969; Staudte and Pette 1972; Pette and Hofer 1980). Moreover, discriminative activity ratios of selected enzymes have been shown to represent highly sensitive indicators of metabolic specialization and have proven to be extremely useful for delineating various "metabolic types" in comparative studies on whole muscles (Bass et al. 1969, 1970; Staudte and Pette 1972).

The development of a technique for single fiber dissection made it possible to determine enzyme activity profiles microbiochemically in histochemically typed muscle fibers (Essén et al. 1975). As a result of the work of Lowry and colleagues (see Lowry and Passonneau 1972), the sensitivity of enzymatic methods of analysis has been increased in such a way that enzyme activity profiles can now be measured in fragments of single muscle fibers. Several studies using single fibers microdissected from rabbit (Spamer and Pette 1977, 1979, 1980; Buchegger et al. 1984; Pette and Spamer 1986), rat (Hintz et al. 1980, 1982, 1984a,b; Takekura and Yoshioka 1987), guinea pig (Hirsch et al. 1982), and human (Lowry et al. 1978, 1980; Chi et al. 1983) muscles have shown that great differences exist in the enzyme activity profiles of individual fibers within both the fast and slow fiber populations. The range of differences greatly exceeds the variations observed by qualitative enzyme histochemistry and demonstrates a metabolic continuum, a notion proposed as early as 1971 by Guth and Yellin.

Discriminative enzyme activity ratios (Bass et al. 1969) may be used to delineate fast- and slow-twitch fiber populations (Fig. 4). Such ratios are formed by referring enzyme activities of anaerobic pathways to those of aerobic pathways (e.g., phosphofructokinase/citrate synthase, lactate dehydrogenase/malate dehydrogenase, glyceraldehydephosphate dehydrogenase/3-hydroxyacyl-CoA dehydrogenase) or by comparing the enzyme activities of nonequilibrium reactions (e.g., hexokinase/phosphofructokinase, phosphofructokinase/fructose 1,6-bisphosphatase; Pette and Spamer 1986). The activity ratio lactate dehydrogenase/adenylate kinase has also been used to successfully delineate fast and slow fiber populations (Lowry et al. 1978; Hintz et al. 1982, 1984a, b). Finally, the microphotometrically assessed activity ratio of glycerolphosphate dehydrogenase/succinate dehydrogenase has been shown to be of a suitable magnitude for delineating metabolically distinct fiber subgroups by quantitative histochemistry (Reichmann and Pette 1982, 1984; Spurway and Rowlerson 1989).

On the basis of this type of analysis, myosin-based functional properties seem to correlate, to some degree, with metabolic profiles. However, there is pronounced metabolic heterogeneity within each group and, therefore, this relationship does not appear to extend beyond the two major groups of fast-twitch and slow-twitch fibers (Reichmann and Pette 1982; Hintz et al. 1984b; Nemeth and Turk 1984; Pette 1985; Pette and Spamer 1986). This implies that enzyme activity measurements performed on pooled fibers (separated on the basis of mATPase histochemistry) give misleading results (Essén-Gustavsson and Henriksson 1984) because their metabolic heterogeneity may be masked.

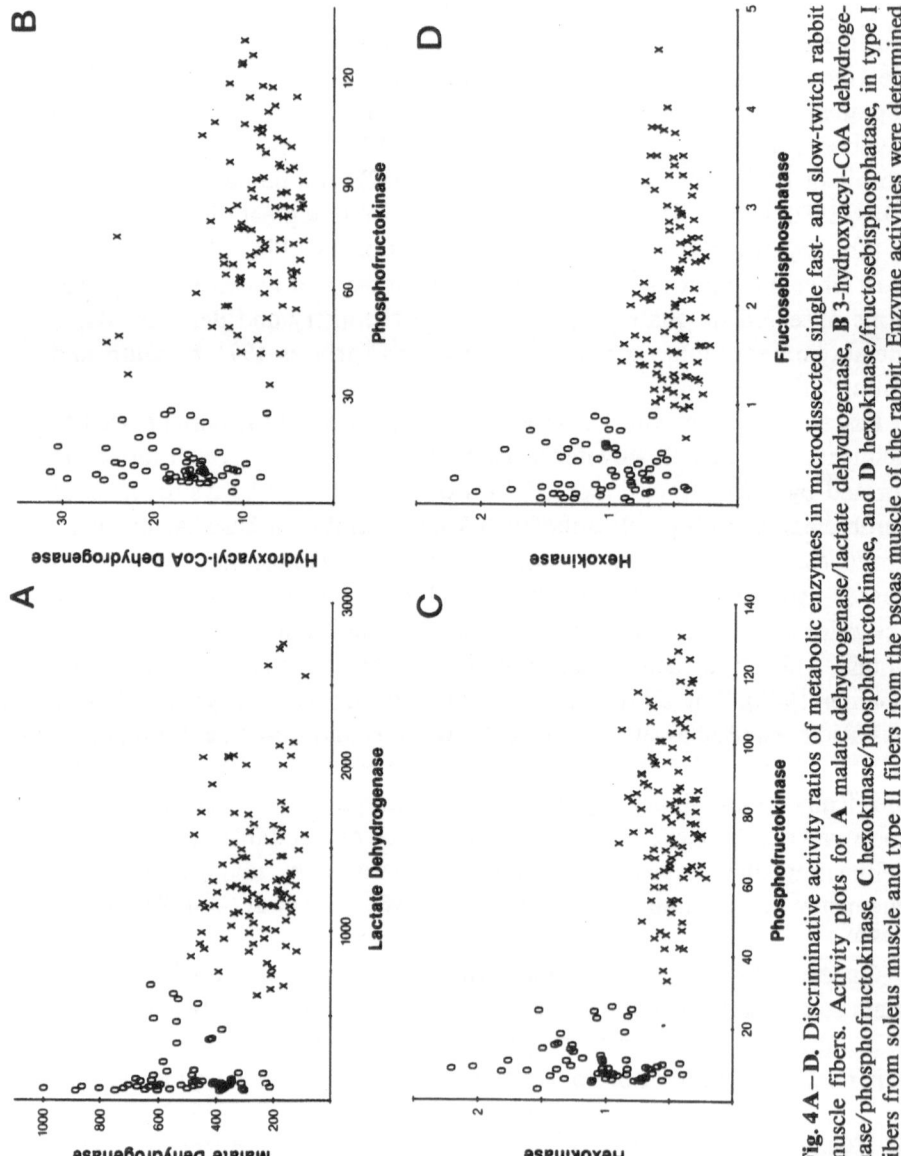

Fig. 4A–D. Discriminative activity ratios of metabolic enzymes in microdissected single fast- and slow-twitch rabbit muscle fibers. Activity plots for **A** malate dehydrogenase/lactate dehydrogenase, **B** 3-hydroxyacyl-CoA dehydrogenase/phosphofructokinase, **C** hexokinase/phosphofructokinase, and **D** hexokinase/fructosebisphosphatase, in type I fibers from soleus muscle and type II fibers from the psoas muscle of the rabbit. Enzyme activities were determined in fragments of the same fibers and are given as micromoles per minute per gram of muscle. *O*, type I fibers; *X*, type II fibers. (From Pette and Spamer 1986)

The inability to differentiate metabolically between fast fiber subtypes also applies to lactate dehydrogenase isozyme patterns (Leberer and Pette 1984). Type I and type II fibers can be separated because they contain distinct isozyme patterns, with a predominance of H-subunit-based isozymes in type I- and M-subunit-based isozymes in type II. However, no clear-cut differences exist between the isozyme patterns of type IIA and type IIB fibers. Likewise, fiber types I and II seem to be distinguished by differences in carbonic anhydrase isozyme content. Immunohistochemical studies of the distribution of the three skeletal muscle isozymes of carbonic anhydrase demonstrate that isozyme III predominates in type I fibers of rat and human muscles (Väänänen et al. 1982, 1986; Shima et al. 1983; Jeffery et al. 1986). Although some reports suggest that this isozyme may not be confined exclusively to type I fibers (Jeffery et al. 1987; Frémont et al. 1988), recent findings strongly support the notion that the expression of carbonic anhydrase III is correlated with that of the slow myosin heavy chain isoform (Jeffery et al. 1990).

Myoglobin, the oxygen-binding hemoprotein which contributes to the macroscopic differences in color between muscles (white and red muscles), has also been used to delineate major fiber types in human (Alway et al. 1988) and in rat muscle (Krenács et al. 1989). In rat muscle, FOG fibers exhibit the highest myoglobin content, SO fibers an intermediate content, and FG fibers the lowest content (Krenács et al. 1989). Pronounced increases in myoglobin content in rat fast-twitch muscles occur following endurance training (Pattengale and Holloszy 1967) and chronic low-frequency electrostimulation (Kaufmann et al. 1989). However, it has been known for many years (Burleigh and Schimke 1969) that the myoglobin content and the activity levels of key aerobic oxidative metabolic enzymes do not necessarily parallel each other in some species.

Microphotometric studies reveal a weak correlation between myoglobin content and succinate dehydrogenase activity in single fibers of mouse muscles (van der Laarse et al. 1985). Likewise, studies on human muscles (Jansson et al. 1982; Svedenhag et al. 1983) and analyses of specific human muscle fibers (Nemeth and Lowry 1984) indicate that type I and type II fibers show only small differences in myoglobin content. These studies also show that training-induced changes in activities of aerobic oxidative metabolic enzymes are not paralleled by altered myoglobin levels. Therefore, at least in the human, myoglobin appears to be a poor marker for mATPase-based fiber types and for oxidative potential.

The metabolic difference between fast and slow muscles is less pronounced in the newborn, but increases during postnatal development (Bass et al. 1970; Margreth et al. 1970; Hudlická et al. 1973; Dalrymple et al. 1974; Zuurveld et al. 1985). The increasing metabolic heterogeneity during

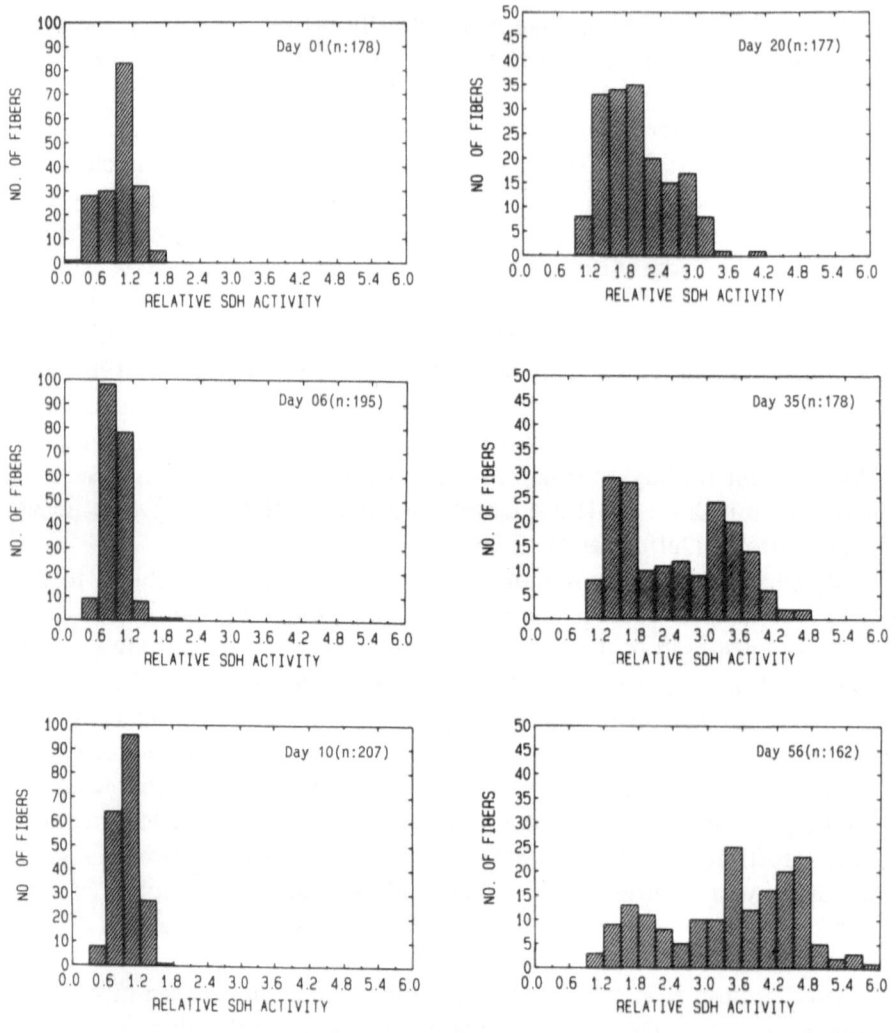

Fig. 5. Postnatal development of metabolic fiber diversity as illustrated by relative succinate dehydrogenase (*SDH*) activities in the fiber populations of tibialis anterior muscles from 1-, 6-, 10-, 20-, 35-, and 56-day-old rats. Enzyme activities were measured microphotometrically (Pette 1981) on cross sections from composite blocks containing fiber bundles of tibialis anterior muscles and adult rat psoas muscle. The activities determined in the tibialis anterior muscles of different ages were referred to the same reference fiber from the psoas muscle. Increases in enzyme activity can, therefore, be compared between the different developmental stages. (B. Huber and D. Pette, unpublished)

postnatal development is especially evident at the single fiber level (Dangain et al. 1987; Nemeth et al. 1989; Fig. 5). It most likely results from changing usage patterns and reflects different properties of individual motor units.

To date, it appears that singly innervated mammalian muscle fibers within a given motor unit are biochemically similar or identical. This assumption is based upon the original qualitative enzyme histochemical observations of Edström and Kugelberg (1968), Kugelberg (1973), and Burke et al. (1971, 1973, 1974). Semiquantitative and quantitative analyses of enzyme activities also indicate less variability between fibers within a motor unit than between fibers of different motor units (Kugelberg and Lindegren 1979; Nemeth et al. 1981; Nemeth and Turk 1984; Vetter et al. 1984; Hamm et al. 1988; Martin et al. 1988a, b; Nemeth and Wilkinson 1990). However, data based upon larger sample sizes and microphotometrically evaluated enzyme histochemistry suggest that a greater variability may exist between fibers within a motor unit (Martin et al. 1988a, b). Local factors (interfascicular and intramuscular location of fibers composing a given motor unit, capillarization, availability of oxygen and metabolites, active and passive stretch, temperature gradients, etc.) may modulate the primary pattern dictated by the neural input. As such, nonuniformity may exist, to some degree, between fibers within the same motor unit as well as within a single fiber (Staron and Pette 1987c).

3.2 Myofibrillar Proteins

Qualitative mATPase-based histochemistry indicates a continuum between the three main fiber types (Andersen and Henriksson 1977; Jansson and Kaisser 1977; Jansson et al. 1978; Green et al. 1979; Ingjer 1979; Häggmark et al. 1981; Howald 1982; Staron et al. 1983) and suggests differences in the molecular composition of myosin within different fibers. Indeed, isoforms of myosin and other myofibrillar proteins have been found in whole muscle preparations. It is clear, however, that the complexity of skeletal muscle as a tissue can only be understood by investigating the myofibrillar protein composition of individual muscle fibers.

3.2.1 Myosin

Myosin is an asymmetric, hexameric protein consisting of four light chains (approximately 20 kDa each) and two heavy chains (approximately 200 kDa each; Gershman et al. 1969; Lowey et al. 1969; Gazith et al. 1970; Dow and Stracher 1971; Lowey and Risby 1971; Weeds and Lowey 1971). This ubiquitous protein, found in both muscle and nonmuscle cells, makes up the largest part of the contractile apparatus in skeletal muscle fibers (Obinata et al. 1981; Murakami and Uchida 1985; Ohtsuki et al. 1986). The two myosin heavy chains intertwine at the carboxyl terminus forming an

α-helical coiled coil (the rod). The amino terminus of each heavy chain forms a globular head region such that each holomyosin contains two heads. The rod is responsible for the aggregation of myosin molecules into bipolar thick filaments, whereas the head region is the site of ATPase activity and actin binding.

3.2.1.1 Myosin Light Chains

The four myosin light chains (for review, see Matsuda 1983) are associated with the two myosin heads. The bound light chains consist of a pair of phosphorylatable (DTNB[1], P or regulatory) light chains (LC2) and a pair of alkali (essential, catalytic) light chains (LC1 and/or LC3). One alkali light chain subunit and one P light chain subunit are associated with the head region of each myosin heavy chain. Although the role of the light chains is poorly understood (Wagner and Giniger 1981), their location near the hinge region (head/rod junction; Flicker et al. 1983; Waller and Lowey 1985; Winkelmann and Lowey 1986; Katoh and Lowey 1989) suggests that they may be involved in modulating interactions between myosin and actin (Wagner and Weeds 1977; Moss et al. 1982; Margossian et al. 1983; Trayer and Trayer 1985; Schaub et al. 1986). Evidence suggests a correlation between the maximum shortening velocity and the alkali light chain ratio (LC1f/LC3f) in type IIB fibers from different rabbit muscles (Sweeney et al. 1988; Greaser et al. 1988; Moss et al. 1990). Likewise, the phosphorylatable light chain appears to have an influence on the mechanical properties. Removal or addition of this light chain decreases or increases, respectively, the maximum velocity of shortening in skinned muscle fibers (Moss et al. 1982).

Adult mammalian fast muscle contains two distinct alkali light chains (LC1f and LC3f) and a phosphorylatable light chain (LC2f). Adult mammalian slow muscle contains the alkali light chain LC1s and a phosphorylatable light chain LC2s (Locker and Hagyard 1967; Stracher 1969; Lowey and Risby 1971; Sarkar et al. 1971; Weeds and Lowey 1971; Sréter et al. 1972; Weeds 1976). Light chains LC1f and LC3f are derived from a single gene by a combined process of differential transcription and RNA splicing (Nabeshima et al. 1984; Periasamy et al. 1984; Robert et al. 1984; Strehler et al. 1985; Barton and Buckingham 1985; Andreadis et al. 1987). As judged from electrophoretic and peptide cleavage pattern analyses, the slow myosin light chains LC1s and LC2s appear to be identical with the cardiac (ventricular) myosin light chains LC1v and LC2v (Lowey and Risby 1971; Sarkar et al. 1971; Weeds 1976; Whalen et al. 1978; Dalla

---

[1] DTNB = 5,5'-dithio bis-(2 nitrobenzoic acid)

Libera et al. 1979; Margreth et al. 1980a; Dalla Libera 1988). In the mouse (Barton et al. 1985b) and rat (Periasamy et al. 1989) both slow skeletal myosin LC1s and cardiac LC1v are encoded by a single gene. Peptide pattern and amino acid sequence analyses strongly indicate the identity of the phosphorylatable light chains LC2s and LC2v in slow and cardiac (ventricle) muscles of rabbit and chicken (Collins et al. 1986; Dalla Libera 1988).

Numerous variants of both fast and slow light chains have been described. Variants of the fast light chains LC1f and LC2f exist in the superfast contracting fibers of jaw-closing muscles (Rowlerson et al. 1981; Shelton et al. 1988). In certain species, two slow alkali light chains (LC1sa and LC1sb) are present (Weeds 1976; Schachat et al. 1980; Pinter et al. 1981; Biral et al. 1982; Zimmermann and Starzinski-Powitz 1989); LC1sa is slightly heavier than LC1sb, and marked differences exist between the isoelectric point (pI) values of these two variants. Pronounced differences also exist in the LC1sa pI values of various mammalian species (Carraro et al. 1981a). In addition, the relative amounts of LC1sa and LC1sb vary between homologous muscles of different species (Carraro et al. 1981a), between different muscles within a species (Schachat et al. 1980; Margreth et al. 1980b; Carraro et al. 1981a; Pinter et al. 1981; Biral et al. 1982; Salviati et al. 1982, 1983; Mabuchi et al. 1984), and even between fibers of the same histochemical type within a given muscle (Salviati et al. 1983; Staron and Pette 1987a, b). These observations could indicate the existence of an even greater number of LC1s isoforms.

Indeed, the LC1sb of the masseter muscle is antigenically different from the LC1sb of the slow-twitch limb muscle in rabbit (Biral et al. 1982). Also, two forms of the P-light chain (LC2s and LC2s') have been found in slow fibers of rabbit and human muscles (Westwood et al. 1984; Houston et al. 1985, 1987; Pernelle et al. 1988). With the use of an improved separation technique, an increased number of variants for most of the myosin light chains have been reported for rabbit muscles, i.e., four for LC1s, three for LC2s, four for LC2f, and three for LC3f (Pernelle et al. 1986). Evidence also exists that LC1f occurs as three variants in avian fast-twitch muscle (Rushbrook et al. 1988a).

An embryonic, nonphosphorylatable light chain ($LC1_{emb}$) has been demonstrated in rat (Whalen et al. 1978, 1982) and in human muscle (Strohman et al. 1983; Biral et al. 1984). In the rat, $LC1_{emb}$ appears to be identical with fetal LC1v (Price et al. 1980; Cummins and Price 1980; Whalen and Sell 1980) and adult atrial LC1 ($LC1_a$; Long et al. 1977; Syrovy et al. 1979). In the mouse, there is a single gene for $LC1_a$ and the embryonic light chain $LC1_{emb}$. In addition, the specific mRNAs present in adult atrial muscle and fetal skeletal muscle are indistinguishable (Barton et al. 1985 a, b). Studies with a cDNA isolated from fetal human muscle also indicate that LC_{emb} and adult $LC1_a$ are encoded by a single gene (Seidel et al. 1988). Adding to this ever growing list of light chains is the expression of a phosphorylatable embryonic light chain ($LC2_{emb}$) during fetal development in human skeletal muscle (Pons et al. 1987).

Microelectrophoretic analyses on single fibers show that fast myosin light chains are present in fast-twitch fibers and slow myosin light chains are present in slow-twitch fibers (Weeds et al. 1975; Pette and Schnez 1977 a; Pette et al. 1979; Schachat et al. 1980; Billeter et al. 1981 a; Ishiura et al. 1981; Mikawa et al. 1981; Young and Davey 1981; Mabuchi et al. 1982; Salviati et al. 1982, 1983; Baumann et al. 1984). However, evidence indicates that a particular set of light chain isoforms need not exclusively combine with a specific myosin heavy chain. Coexisting fast and slow light chains have been observed in fibers containing either fast or slow heavy chains. Some type I fibers in human muscle express, in addition to slow light chains, various combinations of fast light chains (Billeter et al. 1981 a; Ishiura et al. 1981). Furthermore, rat, rabbit, and bovine muscle fibers histochemically classified as type II A contain fast light chains and, additionally, either LC1sb (Salviati et al. 1982; Young and Davey 1981) or LC1sb and LC2s (Mizusawa et al. 1982; Staron and Pette 1986, 1987 a). These fibers do not express both fast and slow heavy chains and, therefore, do not belong to the C fiber population.

3.2.1.2 Myosin Heavy Chains

Like the light chains, myosin heavy chains are encoded by a highly conserved multigene family (Nguyen et al. 1982; Wydro et al. 1983; Buckingham et al. 1986; Mahdavi et al. 1986; Emerson 1987). Early investigations distinguished two different skeletal muscle myosins, fast and slow (e.g., Locker and Hagyard 1967; Lowey and Risby 1971; Nakamura et al. 1971; Sarkar et al. 1971; Jean et al. 1973; Arndt and Pepe 1975; Balint et al. 1975). The detection of antigenic differences between myosins (Gröschel-Stewart and Doniach 1969; Masaki 1974; Arndt and Pepe 1975; Bruggmann and Jenny 1975) initiated immunohistochemical analyses of muscle fibers with antibodies raised against myosins from muscles with a predominance of either fast or slow fiber types (e.g., Masaki 1974; Arndt and Pepe 1975; Gauthier and Lowey 1977, 1979; Lutz et al. 1978, 1979; Billeter et al. 1980). Affinity-purified polyclonal antibodies (against subfragment S1 and light chain LC3f) revealed differences in myosin composition in the fast fiber population of rat muscles (Gauthier and Lowey 1977, 1979). Subsequently, immunohistochemical studies with polyclonal antibodies against specific myosins uncovered differences between the mATPase-

Table 1. Myosin heavy chain isoforms identified in mammalian skeletal muscles

Designation	Nomenclature	Distribution[a]
Embryonic	HC_{emb}	Myotubes, extraocular muscle fibers, intrafusal fibers, regenerating fibers
Neonatal	HC_{neo}	Neonatal muscles, extraocular muscles, masseter, intrafusal fibers, regenerating fibers
Fast-twitch	HCIIb	IIB, IIBD and IIAB fibers
Fast-twitch	HCIIa	IIA, IIAB, IIDA, IIC, IC fibers
Fast-twitch	HCIId (HC2x)	IID (2X), IIBD, IIDA fibers
Fast-twitch	HC_{eom}	Super-fast fibers in extraocular muscles
Fast-twitch	HCIIm	Super-fast fibers in muscles derived from the 1st branchial arch (identified in carnivores and primates)
Slow-twitch	HCI (βHC_{card})	Type I, IC, IIC fibers
Slow-tonic	HCI_{ton}	Slow-tonic fibers in extraocular and tensor tympani muscles and intrafusal fibers

[a] For references see text.

based fiber types IIA and IIB (Lutz et al. 1979; Pierobon-Bormioli et al. 1981). These observations were confirmed and extended by the use of antibodies against specific myosin heavy chains (Danieli-Betto et al. 1986; Schantz and Dhoot 1987).

Unique myosin heavy chains have been detected immunohistochemically in fibers of specific muscles, e.g., extraocular muscles (Wieczorek et al. 1985; Sartore et al. 1987), and muscles derived from the first branchial arch (the masseter, temporalis, pterygoideus, tensor veli palatini, tensor tympani, digastricus anterior, and mylohyoideus muscles; Mascarello et al. 1982, 1983; Rowlerson et al. 1981, 1983; Shelton et al. 1988).

To date, a total of nine distinct myosin heavy chain isoforms have been identified in adult mammalian skeletal muscles (Table 1). The improvement in resolution of protein analytical methods, the use of recombinant DNA techniques, and the widening of the range of muscles and species investigated will, most likely, contribute to increasing the number of defined myosin heavy chain isoforms. According to Robbins et al. (1986), the number of genes within the myosin heavy chain family in the chicken is thought to be 31. A schematic illustration of the electrophoretic mobilities of the major myosin heavy chain isoforms delineated to date in rat skeletal muscle is given in Fig. 6.

The nine defined myosin heavy chains in adult mammalian muscle fibers include five fast, two slow, and two "developmental" heavy chains (Table 1). Two of the fast heavy chains, HCIIa and HCIIb, exist in many different species (Dalla Libera et al. 1980; Dalla Libera 1981; Billeter et al. 1981a; Pierobon-Bormioli et al. 1981; Zweig 1981; Young and Davey 1981; Billeter

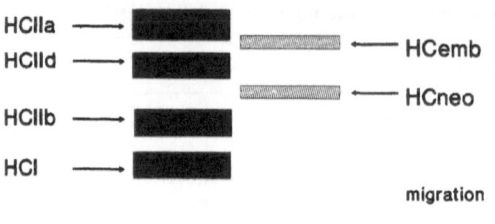

Fig. 6. Schematic illustration of the electrophoretic mobilities of embryonic (*HC_emb*), neonatal (*HC_neo*), adult fast (*HCIIa, HCIId, HCIIb*), and slow myosin heavy chain (*HCI*) isoforms of rat skeletal muscle (from Termin et al. 1989a)

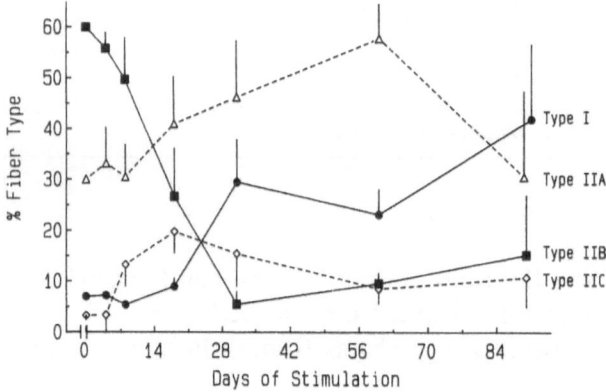

Fig. 7. Time course of changes in histochemically (myofibrillar ATPase activity) classified fiber types in chronically low-frequency stimulated (10 Hz, 12 h/day) extensor digitorum longus muscles of the rabbit. Values represent means ± standard error of mean. (Data from Maier et al. 1988a)

et al. 1982; Salviati et al. 1982; Young 1982; Danieli Betto et al. 1986; Perrie and Bumford 1986; Staron and Pette 1987b; Termin et al. 1989a, b). Myosin HCIIb is present in type IIB fibers, and myosin HCIIa in type IIA fibers. These fast myosin heavy chains may also coexist in the same fiber (type IIAB fibers). In addition, HCIIa may coexist with the slow heavy chain HCI to give type IIC (a higher proportion of HCIIa than HCI) or type IC fibers (a lower proportion of HCIIa than HCI; Staron and Pette 1986, 1987a, b). The percentage of hybrid fibers, i.e., fibers coexpressing two or more myosin heavy chains, appears to vary in different species and increases under conditions of induced fiber transformation (Staron et al. 1987; Maier et al. 1988a; Fig. 7).

An additional fast fiber type (type 2X) has been identified in muscles of mouse, rat, and guinea pig by using monoclonal antibodies (Schiaffino et al. 1985, 1986c, 1988a, 1989, 1990; Gorza 1990). This fiber type contains a specific myosin heavy chain (HC2x) which has been identified by immunoblot analysis on whole muscle homogenates by Schiaffino et al.

Table 2. Densitometrically evaluated percentage distributions of myosin heavy chain isoforms in various muscles of adult Wistar rats

Muscle	MHCI (%)	MHCIIb (%)	MHCIId (%)	MHCIIa (%)
Levator ani	0	100	0	0
Vastus (superficial)	0	98	2	0
Gastrocnemius (superficial)	0	88	12	0
Psoas (superficial)	0	88	12	0
Plantaris	3	51	46	0
Tibialis (anterior)	0	77	19	4
Psoas (deep)	0	71	23	6
Tongue musculature	0	20	72	8
Extraocular muscles	0	24	67	9
Mylohyoideus	0	56	34	10
Gastrocnemius (deep)	10	26	50	14
Soleus	100	0	0	0
Masseter (deep)	0	7	74	19
Vastus (deep)	6	23	50	21
Diaphragm	14	1	53	32

Data are from Bär and Pette (1988) according to Termin et al. (1989a). MHC, myosin heavy chain

(1989, 1990). The HC2x isoform contains an epitope common to HCIIb and another epitope common to all type II heavy chains. Another fast myosin heavy chain isoform has been independently identified in various skeletal muscle of the rat (Table 2) by an improved gradient gel electrophoretic technique (Bär and Pette 1988). Because of its abundance in muscles specialized for sustained activity, especially the diaphragm, this isoform was tentatively named HCIId (Bär and Pette 1988; Termin et al. 1989a,b, 1990). The electrophoretic mobility of HCIId is intermediate between that of HCIIa and HCIIb (Fig. 6). It remains to be seen whether or not some of the electrophoretically separated myosin heavy chain bands designated as HCIIa in the literature have been erroneously assigned and in fact represent the newly detected heavy chain HCIId. Although Schiaffino et al. (1989) were not able to electrophoretically separate HC2x from HCIIa, inspection of their published immunoblot analyses suggests that HC2x and HCIId are identical. This assumption is supported by independent observations indicating that HC2x/HCIId increase in chronically electrostimulated muscles (Schiaffino et al. 1988a, 1989; Bär and Pette 1988; Termin et al. 1989a,b, 1990).

Electrophoretic analyses of single, histochemically classified fibers isolated from hindlimb muscles and diaphragm of the rat made it possible to assign HCIId to a specific fast fiber population (Termin et al. 1989a,b,

1990). With the exception of slight differences in the reactivity of mATPase after acid preincubation (pH 4.6), fiber types IID and IIB are almost indistinguishable in rat skeletal muscle (Termin et al. 1989a, b, 1990). However, type 2X/IID fibers can be distinguished from type IIB fibers by the higher alkaline stability of their mATPase activity after formaldehyde pretreatment (Schiaffino et al. 1986c, 1989; Gorza 1990). Single fiber analyses performed on rat extensor digitorum longus and diaphragm reveal that HCIId and HCIIb may coexist in variable proportions in individual fibers, i.e., type IIBD fibers (Termin et al. 1989a, b, 1990).

As shown in single fibers from rat muscles undergoing fast-to-slow transitions during chronic low-frequency stimulation, HCIId may coexist with HCIIa in various proportions and, in a few fibers, also with minute amounts of HCI (Termin et al. 1989b, 1990). Furthermore, it has been shown that HC2x represents the predominant myosin heavy chain isoform in rat soleus muscle undergoing slow-to-fast transitions during phasic, high-frequency electrostimulation (Schiaffino et al. 1988a, 1989). The shortening velocity of the transformed soleus muscle is intermediate between that of the normal soleus (predominance of type I fibers) and that of the fast extensor digitorum longus muscle (predominance of type IIB fibers), thus indicating that myosin HC2x/IId is functionally intermediate between HCIIb and HCI (Schiaffino et al. 1988a, 1990). This is in agreement with the appearance of HCIId intermediately between decay of HCIIb and increase in HCIIa during the stimulation-induced myosin heavy chain transitions in rat fast-twitch muscle (Termin et al. 1989b, 1990).

Two additional fast myosin heavy chains are expressed in specific muscles. A unique fast myosin heavy chain isoform (HCIIm) is present in jaw-closing muscles and in the tensor tympani muscle of the cat and dog (Rowlerson et al. 1981; Mascarello et al. 1982, 1983; Hoh et al. 1988), and a specific fast myosin heavy chain isoform, HC_{eom}, exists in super-fast contracting fibers of the extraocular muscles (Wieczorek et al. 1985; Sartore et al. 1987).

Slow-twitch fibers contain the slow myosin heavy chain isoform HCI (Weeds and Burridge 1975; Gauthier and Lowey 1979; Pierobon-Bormioli et al. 1980) which corresponds to the cardiac heavy chain βHC_{card} (Lompré et al. 1984). An additional slow heavy chain isoform, HCI_{ton}, is associated with slow tonic muscle fibers present in extraocular muscles, the tensor tympani muscle, and intrafusal fibers (Pierobon-Bormioli et al. 1980; Mascarello et al. 1982, 1983; Sartore et al. 1987).

Two specific myosin heavy chain isoforms, the embryonic (HC_{emb}) and neonatal (HC_{neo}), are expressed in developing mammalian skeletal muscles (Sréter et al. 1975; Whalen et al. 1979, 1981; Rushbrook and Stracher 1979; Fitzsimons and Hoh 1981; Bader et al. 1982; Bandman et al. 1982b;

Carraro and Catani 1983; Gambke et al. 1983; Lowey et al. 1983; Lyons et al. 1983; Butler-Browne and Whalen 1984; Gambke and Rubinstein 1984; Stockdale and Miller 1987; Weydert et al. 1987). A second embryonic heavy chain isoform exists in chicken muscle (Lowey et al. 1986; Crow 1987; Gauthier 1987; Hofmann et al. 1988). Recently, it has been demonstrated that these developmental heavy chain isoforms are also expressed in specific adult mammalian muscles, e.g., extraocular muscles (Wieczorek et al. 1985; Sartore et al. 1987), murine masseter muscle (d'Albis et al. 1986), regenerating muscle (Sartore et al. 1982; Carraro et al. 1983; Maier et al. 1986a; Cerny and Bandman 1987; d'Albis et al. 1988; Maier et al. 1988a), denervated muscle (Cerney and Bandman 1987; Schiaffino et al. 1988b; Biral et al. 1989), dystrophic muscle (Bandman 1985a; Schiaffino et al. 1986a; Rushbrook et al. 1987), and intrafusal fibers (Rowlerson et al. 1985; Schiaffino et al. 1986a; Maier et al. 1988b; Kucera and Walro 1989; Pedrosa et al. 1989), and in rhabdomyosarcomas (Schiaffino et al. 1986b).

The number of myosin heavy chain isoforms continues to increase. Electrophoretic studies have recently delineated six different myosin heavy chain isoforms during embryonic development of avian pectoralis muscle (Hofmann et al. 1988). At least four heavy chain species have recently been identified at the protein level in adult fast avian muscle (Rushbrook et al. 1988b, 1990). In addition, possible myosin heavy chain variants may exist due to posttranslational modification (Bandman et al. 1982a).

Taken together, a variety of fiber types can be distinguished in a given muscle by mATPase-based histochemistry, immunohistochemistry or myosin heavy chain analyses. Nevertheless, the number of fiber types is less than the total number of fibers in a given muscle. Assuming uniformity of the motor unit, the number of fiber types should equal the number of motor units. This implies that the phenotypic expression of adult muscle fibers is primarily dictated by the neuromuscular activity pattern (for reviews see Jolesz and Şréter 1981; Salmons and Henriksson 1981; Pette and Vrbová 1985). Indeed, all fibers composing a specific motor unit appear to contain immunohistochemically identical myosin (Gauthier et al. 1983).

The significance of neural control on phenotypic expression of muscle fibers has been illustrated by the use of chronic stimulation (for reviews see Pette and Vrbová 1985; Pette 1986, 1990). However, only a few studies have addressed this relationship with regard to myosin isoforms and fiber types. In the rabbit, chronic low-frequency stimulation induces pronounced fast-to-slow transitions in both myosin isoforms and fiber types. At the myosin level, the fast heavy chain isoforms HCIIb and HCIIa are progressively replaced by the slow heavy chain HCI. The extent of the transformation is dependent upon both the pattern and duration of stimulation.

Intermittent 10 Hz stimulation, i.e., 8 h/day (Mabuchi et al. 1982) or 12 h/day (Maier et al. 1988a) results in less extensive fast-to-slow transitions than continuous (24 h/day) 10 Hz stimulation (Staron et al. 1987). As shown by single fiber analyses and mATPase-based histochemistry, these changes in rabbit muscle correspond to fiber type transitions in the following sequence: IIB→IIAB→IIA→IIC→IC→I (Staron et al. 1987; Maier et al. 1988a; Pette 1990; Fig. 7). Restoration of the fast fiber population and fast-twitch muscle characteristics occurs following cessation of chronic low-frequency stimulation (Eisenberg et al. 1984; Kirschbaum and Pette 1988; Kirschbaum et al. 1989b, 1990a; Brown et al. 1989). In the rat, stimulation-induced transitions in myosin heavy chain expression remain restricted to exchanges within the fast isoforms (Termin et al. 1989b, 1990). In the euthyroid rat, the isoform exchange occurs in the following order: HCIIb→HCIId→HCIIa. Because thyroid hormone counteracts the effects of increased neuromuscular activity as induced by chronic low-frequency stimulation, the range of myosin heavy chain transitions can be extended in the hypothyroid animal (Kirschbaum et al. 1990b). Under hypothyroid conditions, the stimulation-induced transitions occur in the following order: HCIIb→HCIId→HCIIa→HCI.

These changes extend the original notion of fiber type transitions (Jansson et al. 1978) and verify the dynamic nature of muscle fibers (Guth and Yellin 1971). Indeed, the rapid responsiveness of muscle fibers is illustrated by pronounced changes in specific myosin heavy chain mRNAs within hours of initiation or cessation of chronic low-frequency stimulation (Kirschbaum and Pette 1988; Kirschbaum et al. 1989a, b, 1990a; Pette 1990). The ability of muscle fibers to undergo transformation in response to altered functional demands is also evident from numerous training studies using mATPase histochemistry or immunohistochemistry (Andersen and Henriksson 1977; Jansson and Kaijser 1977; Jansson et al. 1978; Green et al. 1979, 1984; Ingjer 1979; Billeter et al. 1981b; Schantz et al. 1982; Luginbuhl et al. 1984; Larsson and Ansved 1985; Simoneau et al. 1985; Schantz and Dhoot 1987; Baumann et al. 1987).

3.2.1.3 Isomyosins

Isomyosins were detected electrophoretically under nondenaturing conditions (d'Albis and Gratzer 1973; Hoh 1975). The formation of isomyosins is the result of the hexameric structure of myosin. Each holomyosin contains two heavy chains in combination with two phosphorylatable (regulatory) light chains and two alkali light chains. Considering a "typical" fast myosin with a pair of identical heavy chains, and assuming the regulatory light chains exist only as the homodimer $(LC2f)_2$, three different light chain-based isomyosins are possible from the three alkali light chain combinations: LC1f homodimer, LC1f/LC3f heterodimer, and LC3f homodimer (Lowey et al. 1979; Lowey 1980). These can be demonstrated electrophoretically under nondenaturing conditions and are designated as FM3, FM2, and FM1, respectively (Hoh et al. 1976; Hoh 1978; Hoh and Yeoh 1979; d'Albis et al. 1979, 1982).

The slow myosin heavy chain (HCI) together with two distinct alkali light chains (LC1sa and LC1sb) should combine to form three isomyosins. Two slow isomyosins (SM1 and SM2) have been separated in adult mammalian muscle (Fitzsimons and Hoh 1981; Pinter et al. 1981; Maréchal et al. 1989). A third slow isomyosin (SM1') has been described in rat hindlimb muscles (Maréchal et al. 1984; Gregory et al. 1986). The existence of three slow isomyosins, SM1, SM2, and SM3, was recently shown in rat and rabbit slow-twitch muscles. Of these, SM1 was identified as the LC1sa homodimer, SM2 as the LC1sa/LC1sb heterodimer, and SM3 as the LC1sb homodimer (Termin and Pette 1990).

An additional fast isomyosin (IM) with an electrophoretic mobility between that of slow isomyosin and that of FM3 has been described (d'Albis et al. 1982, 1986; Fitzsimons and Hoh 1983; Gregory et al. 1986, 1987; Tsika et al. 1987a,b). This intermediate isomyosin (IM) is present in muscles containing type IIA fibers. Therefore, it seems to represent a HCIIa-based isomyosin (Fitzsimons and Hoh 1983; d'Albis et al. 1986). Its light chain complement varies depending on the muscle. The IM from soleus muscles of various small mammals contains LC1f and LC1s together with LC2f (Fitzsimons and Hoh 1983; Maréchal et al. 1989), whereas IM from fast-twitch rat muscle contains LC1f and LC2f homodimers (Fitzsimons and Hoh 1983; d'Albis et al. 1986).

In addition, special fast and slow isomyosins have been reported for the temporalis muscle in the cat (Hoh et al. 1988; Shelton et al. 1988) and for the extraocular muscles in rat (Wieczorek et al. 1985).

A spectrum of fast heavy chain-based isomyosins has recently been observed by comparing electrophoretically separated isomyosins from rat muscles containing predominantly HCIIb, HCIId, or HCIIa (Termin and Pette 1990). For each heavy chain isoform three light chain-based isomyosins can be distinguished, giving a total of nine fast isomyosins in adult fast-twitch muscles of the rat. These have been named FM1b, FM2b, FM3b, FM1d, FM2d, FM3d, FM1a, FM2a, and FM3a. In view of these findings, we suggest the above nomenclature indicating the heavy chain complement of the respective isomyosins. The HCIIb-based isomyosins display the highest electrophoretic mobilities, the HCIId-based are intermediate, and the HCIIa-based are the slowest (Termin and Pette 1990). The various FM1-type isomyosins seem to represent L3f homodimers, the

FM2-type isomyosins LC1f/LC3f heterodimers, and the FM3-type isomyosins LC1f homodimers. It is remarkable that among the HCIIa-based isomyosins FM1a and FM2a are minor forms. The major form is FM3a, which indicates that HCIIa is predominantly associated with LC1f. It appears most likely that FM3a and probably also FM3d are identical with the previously described IM.

Various isomyosins have also been observed in developing muscle. Their electrophoretic mobilities are higher than those of the fast-type isomyosins in adult muscle. As these "developmental" isomyosins appear sequentially during embryonic and perinatal development (Hoh and Yeoh 1979; Whalen et al. 1981; Maréchal et al. 1984; d'Albis et al. 1989a), it is conceivable that they represent myosins containing the embryonic and neonatal heavy chain isoforms HC_{emb} and HC_{neo} in combination with the embryonic LC_{emb} and the adult fast light chain isoforms (Hoh and Yeoh 1979). These "developmental" isomyosins are also found in regenerating muscles (Maréchal et al. 1984; d'Albis et al. 1988, 1989b). In addition, their presence has been observed in selected craniofacial muscles of the adult rat, e.g., the extraocular muscles (Wieczorek et al. 1985), and masseter (Butler-Browne et al. 1988; d'Albis et al. 1989a; Termin and Pette 1990).

Three major neonatal isomyosins can be distinguished, fm1, fm2, and fm3 (Hoh and Yeoh 1979) which have light chain complements homologous to FM1, FM2, and FM3. These neonatal isomyosins can be distinguished from three major embryonic isomyosins displaying lower electrophoretic mobilities (Termin and Pette 1990). In accordance with Hoh (1978), we suggest a nomenclature of eM1, eM2, and eM3 for the embryonic isomyosins and nM1, nM2, and nM3 for the neonatal isomyosins. In contrast to the neonatal isomyosins that contain the adult LC1f and LC3f homo- and heterodimers, the embryonic isomyosins contain $LC1_{emb}$ and LC1f homo- and heterodimers.

With so many heavy and light chain isoforms, a large number of isomyosins are theoretically possible. The possibility of numerous "hybrid" types (e.g., heterodimer heavy chain combinations and various combinations of light chains) further increases the number of potential isomyosins. The existence of heavy chain heterodimers in mammalian skeletal muscle, although suggested (Staron and Pette 1987a, 1990; Pette and Staron 1988), has not been demonstrated. However, the formation of heavy chain heterodimers in cardiac muscle, which had been proposed by Hoh et al. in 1979, has recently been visualized in rat cardiac ventricular myosin (Dechesne et al. 1987).

The coexistence of fast and slow myosins within a single fiber had been suggested after the observation that fast and slow light chains coexist in fibers undergoing transformation from fast to slow (Pette and Schnez

1977b). The use of immunohistochemistry expanded this concept and confirmed the coexpression of fast and slow myosins in developing muscle fibers (Gauthier et al. 1978; Kelly and Rubinstein 1980; Rubinstein and Kelly 1980; Narusawa et al. 1987), as well as in adult muscle fibers histochemically identified as C fibers (Rubinstein et al. 1978; Lutz et al. 1979; Billeter et al. 1980, 1981a; Snow et al. 1981; Salviati et al. 1982, 1983; Baumann et al. 1984; Schantz and Dhoot 1987).

Furthermore, single fiber dissection has demonstrated that certain fibers can coexpress two fast heavy chains, e. g., HCIIa and HCIIb (Danieli-Betto et al. 1986; Staron and Pette 1987b; Biral et al. 1988), while others coexpress fast and slow heavy chains, e. g., HCI and HCIIa (Danieli-Betto et al. 1986; Salviati et al. 1986; Staron and Pette 1986, 1987a, 1990; Staron et al. 1987; Biral et al. 1988; Termin et al. 1989a, b, 1990; Gorza 1990). The fibers coexpressing heavy chains HCIIa and HCIIb may compose a large percentage of the fast fiber population in human muscle (Biral et al. 1988). Recent observations on chronically stimulated rat (Termin et al. 1989b, 1990; Schiaffino et al. 1986c, 1989, 1990; Gorza 1990) and rabbit (Maier et al. 1988a) muscles indicate that transforming fibers may contain more than two heavy chains. Thus, up to four different heavy chains can be separated electrophoretically in single, transforming rat muscle fibers (Termin et al. 1989b). Recent immunohistochemical studies on transitory fibers during chicken muscle development have unambiguously shown the presence of different myosin isoforms within myofibrils of the same fiber (Gauthier 1990).

Considering only the heavy chains HCI, HCIIa, and HCIIb, as well as the slow (LC1s and LC2s) and the fast (LC1f, LC2f, LC3f) light chains, a total of 60 isomyosins (Table 3) could exist in adult mammalian muscle (Staron and Pette 1987a, b, 1990; Pette and Staron 1988). Indeed, single fiber analyses have demonstrated the coexistence of the fast and slow light chain isoforms together with heavy chains HCI and HCIIa in histochemically typed IC and IIC fibers of rabbit soleus and tibialis anterior muscles (Staron and Pette 1986, 1987a, b; Staron et al. 1987). Therefore, C fibers may contain 54 theoretically different isomyosins (Table 3). Of these 54 possible isomyosins, one is expressed in type I fibers and three are expressed in type IIA fibers. Six additional fast isomyosins result from the three possible fast light chain combinations with the heavy chain HCIIb homodimer and from the same light chain combinations with the HCIIa/HCIIb heterodimer. Thus, considering only heavy chains HCIIa and HCIIb, a total of nine fast isomyosins are theoretically possible in rabbit muscle fibers (Staron and Pette 1987b, 1990).

It is not known whether the limited number of isomyosins thus far demonstrated is the result of insufficient resolution of the applied methods or

Table 3. Possible combinations between myosin light and heavy chains derived from single fiber studies in the rabbit

Light chain combinations	Heavy chain combinations		
	A	B	C
1 $(LC1s)_2$ $(LC2s)_2$	$(HCI)_2$	(HCI) $(HCIIa)$	$(HCIIa)_2$
2 $(LC1s)$ $(LC1f)$ $(LC2s)_2$	$(HCI)_2$	(HCI) $(HCIIa)$	$(HCIIa)_2$
3 $(LC1f)_2$ $(LC2s)_2$	$(HCI)_2$	(HCI) $(HCIIa)$	$(HCIIa)_2$
4 $(LC1s)_2$ $(LC2s)$ $(LC2f)$	$(HCI)_2$	(HCI) $(HCIIa)$	$(HCIIa)_2$
5 $(LC1s)$ $(LC1f)$ $(LC2s)$ $(LC2f)$	$(HCI)_2$	(HCI) $(HCIIa)$	$(HCIIa)_2$
6 $(LC1f)_2$ $(LC2s)$ $(LC2f)$	$(HCI)_2$	(HCI) $(HCIIa)$	$(HCIIa)_2$
7 $(LC1s)_2$ $(LC2f)_2$	$(HCI)_2$	(HCI) $(HCIIa)$	$(HCIIa)_2$
8 $(LC1s)$ $(LC1f)$ $(LC2f)_2$	$(HCI)_2$	(HCI) $(HCIIa)$	$(HCIIa)_2$
9 $(LC3f)_2$ $(LC2s)$ $(LC2f)$	$(HCI)_2$	(HCI) $(HCIIa)$	$(HCIIa)_2$
10 $(LC3f)_2$ $(LC2s)_2$	$(HCI)_2$	(HCI) $(HCIIa)$	$(HCIIa)_2$
11 $(LC1s)$ $(LC3f)$ $(LC2s)_2$	$(HCI)_2$	(HCI) $(HCIIa)$	$(HCIIa)_2$
12 $(LC1f)$ $(LC3f)$ $(LC2s)_2$	$(HCI)_2$	(HCI) $(HCIIa)$	$(HCIIa)_2$
13 $(LC1f)$ $(LC3f)$ $(LC2s)$ $(LC2f)$	$(HCI)_2$	(HCI) $(HCIIa)$	$(HCIIa)_2$
14 $(LC1s)$ $(LC3f)$ $(LC2f)_2$	$(HCI)_2$	(HCI) $(HCIIa)$	$(HCIIa)_2$
15 $(LC1s)$ $(LC3f)$ $(LC2s)$ $(LC2f)$	$(HCI)_2$	(HCI) $(HCIIa)$	$(HCIIa)_2$
16 $(LC1f)_2$ $(LC2f)_2$	$(HCI)_2$	(HCI) $(HCIIa)$	$(HCIIa)_2$
17 $(LC1f)$ $(LC3f)$ $(LC2f)_2$	$(HCI)_2$	(HCI) $(HCIIa)$	$(HCIIa)_2$
18 $(LC3f)_2$ $(LC2f)_2$	$(HCI)_2$	(HCI) $(HCIIa)$	$(HCIIa)_2$
19 $(LC1f)_2$ $(LC2f)_2$		$(HCIIa)$ $(HCIIb)$	$(HCIIb)_2$
20 $(LC1f)$ $(LC3f)$ $(LC2f)_2$		$(HCIIa)$ $(HCIIb)$	$(HCIIb)_2$
21 $(LC3f)_2$ $(LC2f)_2$		$(HCIIa)$ $(HCIIb)$	$(HCIIb)_2$

Assuming light and heavy chain heterodimers, and not considering the light chain LC1sa/LC1sb heterogeneity, a total of 60 theoretical isomyosins are possible. Type I fibers contain one combination (1A), types IC and IIC contain 54 combinations (1A–C to 18A–C). Type IIA contain three combinations (16C, 17C, 18C). Type IIB fibers also contain three combinations (19C, 20C, 21C). Type IIAB fibers contain nine combinations (16C–21C plus 19B, 20B, 21B). (From Pette and Staron 1988)

if this reflects preferential affinities between specific myosin light and heavy chains. The electrophoretically detected number of isomyosins continues to increase (e.g., Maréchal et al. 1984; Gregory et al. 1986; d'Albis et al. 1988; Termin and Pette 1990).

The number of possible isomyosins in adult mammalian skeletal muscle may be far greater if one considers the additional myosin light and heavy chain variants which have been detected. Moreover, the variability of muscle fibers depends not only upon the presence of particular sets of myosin light and heavy chains, and upon other myofibrillar protein isoforms, but also depends upon the relative concentrations of the individual compo-

Fig. 8. Coexpression of myosin heavy chains HCIIa and HCI in varying ratios in C fibers of the rabbit. *Upper panel*, cross section stained for mATPase (preincubation at pH 4.3) of a chronically stimulated (10 Hz, 24 h/day, 30 days) tibialis anterior muscle of the rabbit. *Lower panel*, tracings of densitometric evaluations of electrophoretically separated myosin heavy chains from fragments of the same fibers numbered in the *upper panel*. *1*, type IIA fiber; *2–5*, C fibers; *6*, type I fiber. (From Pette and Staron 1988)

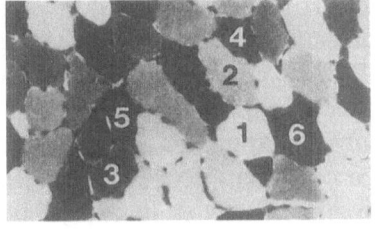

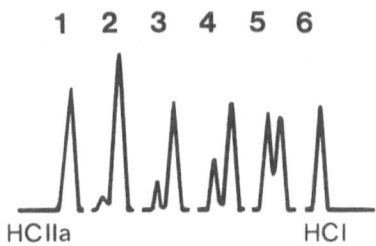

nents. This is best illustrated when examining the C fiber population which expresses a continuum of HCI/HCIIa ratios (Fig. 8; Staron and Pette 1986; Staron et al. 1987; Pette and Staron 1988). Three distinct sets of myofibrils with different proportions of embryonic and neonatal isomyosins were recently shown to coexist within individual fibers of developing chicken muscle by immunofluorescence and immunogold electron microscopy with the use of stage-specific antibodies (Gauthier 1990). Moreover, three classes of thick filaments were delineated in homogenates from developing chicken muscle by immunogold electron microscopy with anti-embryonic and anti-neonatal antibodies. One class of filaments reacted only with the embryonic antibody, a second reacted only with the neonatal-specific antibody, and a third was decorated with both antibodies. Similar results were obtained with filaments prepared from the pectoral muscle of young chicken where the neonatal and adult fast heavy chain isoforms are expressed. Taken together, these observations demonstrate that two myosin isoforms can coexist in vivo within an individual myofibril (Gauthier 1990) and even within an individual thick filament (Taylor and Bandman 1990).

The possibility of nonuniform myosin expression along the length of the muscle fibers further increases the complexity of isomyosin composition. Chronically stimulated and unstimulated, contralateral rabbit tibialis anterior muscles contain fibers with regions of varying myosin expression at the protein (Staron and Pette 1987c) and mRNA levels (unpublished observations). Regional variations in myosin expression have also been demonstrated in denervated rat tibialis anterior and soleus muscles (Schiaffino et al. 1988b). Furthermore, nonuniformity of myosin isoforms exists in normal intrafusal fibers (Maier et al. 1988b; Kucera and Walro 1989; Pedrosa et al. 1989) and may occur along the length of some extrafusal fibers in normal muscles (unpublished observations). The heterogeneous myosin expression along muscle fibers could point to either a loss of nuclear coordination or the existence of regional control by specific nuclear domains (Harris et al. 1989; Pavlath et al. 1989; Hughes and Blau 1990).

3.2.1.4 Correlation Between Myosin Composition and Contractile Properties

Since the work of Bárány (1967), it has generally been assumed that the contractile properties of a given muscle primarily relate to its myosin ATPase activity. This relationship is true for contractile speed (Bárány 1967), but may not be true for other contractile properties, e.g., time to peak of isometric twitch contraction. Additional elements of the muscle fiber, e.g., troponin isoforms or properties of the T tubule system and sarcoplasmic reticulum, may also contribute to these properties. New insights into these relationships have recently arisen from time course studies of low-frequency stimulation-induced fast-to-slow transitions. Changes in Ca^{2+} sequestration, Ca^{2+}-ATPase activity, and the composition of the sarcoplasmic reticulum occur rapidly after the onset of stimulation and correlate with changes in specific contractile properties (Heilmann and Pette 1979; Eisenberg and Salmons 1981; Eisenberg et al. 1984, 1987; Leberer et al. 1986, 1987; Klug et al. 1988; Pette 1989; Simoneau et al. 1989). These adaptations appear to precede changes in protein isoforms of the thick and thin filaments.

There have only been a limited number of investigations dealing with the contractile properties of single fibers. Evidence is available showing correlation between maximum shortening velocity and myosin heavy chain composition. Generally, greater maximum shortening velocities are associated with greater amounts of fast heavy chains, whereas increasing the amounts of the slow heavy chain correlates with slower shortening velocities. Therefore, a continuum of contractile speeds correlates with a continuum of myosin heavy chain compositions. This relationship holds true for adult, developing, and hypokinetic muscles in rat and rabbit (Reiser et al. 1985a, b, 1987a, 1988; Sweeney et al. 1986, 1988; Eddinger and Moss 1987).

Additional support for this myosin heavy chain-contractile speed relationship comes from differences observed within the fast fiber population. Subtle differences in shortening velocity between fiber types IIA and IIB have been demonstrated in rabbit tibialis anterior and psoas muscles

(Sweeney et al. 1986, 1988). Interestingly, differences in unloaded shortening velocity also exist between fibers of the same fast histochemical (mATPase-based) type from two different muscles. The type IIB fibers from rabbit psoas and tibialis anterior muscles contain indistinguishable myosin heavy chains, but differ in their alkali light chain (LC1f/LC3f) ratio (Sweeney et al. 1988). Likewise, Eddinger and Moss (1987) describe two fast-twitch fiber populations in rat diaphragm differing in cross-sectional area and maximum shortening velocities. Although their heavy chain compositions appear to be identical, differences exist in the myosin light chain and regulatory protein content. However, it seems possible that some myosin heavy chain differences may not have been elucidated in this study, particularly since recent evidence suggests the presence of the heavy chain isoforms HCIIb and HCIId, which can only be separated electrophoretically by means of a refined gradient gel system (Bär and Pette 1988; Termin et al. 1989a,b).

3.2.2 Actin

In skeletal muscle, globular actin molecules (G-actin) aggregate to form long, filamentous polymers (F-actin) with a defined polarity. Actin filaments are double-helical arrays of two F-actins. Like myosin, actin is found in both nonmuscle and muscle tissues. However, unlike myosin, actin appears to be highly conserved, with the actin sequence varying little between diverse sources (Elzinga et al. 1973; Vandekerckhove and Weber 1979). The conservative nature of the actin molecule is compatible with its ability to bind to numerous other proteins (Obinata et al. 1981). Still, some isoforms have been found. Muscle actins (cardiac and skeletal α-actins) are distinguishable from nonmuscle and myoblast precursor actins (β- and γ-actins; Whalen et al. 1976; Garrels and Gibson 1976). The two α-actin genes are both expressed at early stages of muscle development, whereas in the adult, skeletal α-actin predominates (Schwartz and Rothblum 1981; Minty et al. 1982; Gunning et al. 1983; Buckingham et al. 1984; Hayward and Schwartz 1986; Minty et al. 1986). As yet, two-dimensional peptide patterns of skeletal muscle α-actins from adult mammalian fiber types (I, IIA, and IIB) appear to be identical (Billeter et al. 1982). However, actin isoforms may arise after modifications during development (Whalen et al. 1976) or under pathologic conditions (Strankfeld and Moskalenko 1987).

3.2.3 Tropomyosin

Tropomyosin and troponin, the two regulatory proteins associated with the actin filament (Potter and Gergely 1974; Leavis and Gergely 1984; Perry

1985; El-Saleh et al. 1986), also display polymorphism. Tropomyosin (TM) is a dimeric protein consisting of two subunits designated α-TM and β-TM (Cummins and Perry 1973, 1974). At the mRNA level, the α-TM subunit exists in at least six different tissue-specific isoforms resulting from alternative splicing of the primary transcript of the α-TM gene (Ruiz-Opazo and Nadal-Ginard 1987; Wieczorek et al. 1988). The β subunits of TM in skeletal and smooth muscle are derived from transcripts of a single gene with a unique promoter by alternative splicing (Libri et al. 1989). In skeletal muscle, three combinations of tropomyosin subunits are possible: α/α and β/β homodimers, and the α/β heterodimer. The α/β-TM subunit ratio appears to be species-specific (Cummins and Perry 1973) and muscle-specific (Roy et al. 1979; Carraro et al. 1981 b; Matsuda et al. 1983).

Two tropomyosin subunit combinations predominate in different rabbit muscles (Bronson and Schachat 1982; Brown and Schachat 1985). The α/α homodimer is prevalent in some specific fast, white muscles of the rabbit (the longissimus dorsi and psoas muscles), whereas the α/β heterodimer is prevalent in the plantaris muscle (defined by these authors as another fast white muscle). In addition, the α/β heterodimer is found in fast red, slow, and mixed muscles (Bronson and Schachat 1982). These various combinations appear to be coexpressed with specific troponin T subunit isoforms (Schachat et al. 1987). Single fiber analyses of several rabbit muscles additionally show a predominance of the β subunit in type I fibers and a predominance of the α subunit in type IIB fibers (Salviati et al. 1982). According to these authors, the α subunit in type I fibers is dissimilar to that in type IIB fibers. Indeed, slow and fast isoforms of the α subunit (α_s-TM and α_f-TM) have been identified (Cummins and Perry 1974; Steinbach et al. 1980; Bronson and Schachat 1982; Salviati et al. 1982, 1983; Heeley et al. 1983, 1985; Kardami et al. 1983). The coexistence of α_f-TM, α_s-TM, and β-TM in normal rabbit soleus muscle, which is predominantly composed of type I fibers, may increase the possible number of homo- and heterodimers, i. e., α_f/α_f, α_s/α_s, α_f/α_s, α_f/β, α_s/β, and β/β (Härtner et al. 1989). Evidence suggests that fast and slow β-TM isoforms may also exist in human fast- and slow-twitch muscle fibers (Salviati et al. 1983).

The complexity of the tropomyosin system may be even greater than that due to two fast and two slow isoforms of both the α and β subunits. Additional tropomyosin subunits have been detected in rabbit (Heeley et al. 1983, 1985) and human (Romero-Herrera et al. 1982) skeletal muscles. High-resolution electrophoretic techniques have recently resolved rabbit skeletal muscle tropomyosin subunits into six components (Pernelle et al. 1986). It is not clear which of these represent phosphorylated or unphosphorylated variants, or whether fiber type-specific distributions exist.

Two-dimensional peptide maps of tropomyosin from histochemically identified human fiber types I, IIA, and IIB indicate possible fiber type-specific differences in both the α and β subunits (Billeter et al. 1981 b). It is also possible that specific tropomyosin isoforms exist in specific muscles. Indeed, type IIM fibers in cat jaw-closer muscles contain only a single form of tropomyosin which is distinct from the tropomyosins present in other fiber types (Rowlerson et al. 1983).

3.2.4 Troponin

The troponin (TN) complex consists of three different subunits in an equimolar stoichiometry, designated TN-T, TN-I, and TN-C (Schaub and Perry 1969; Ebashi 1972; Greaser and Gergely 1973; Potter 1974; Yates and Greaser 1983; Leavis and Gergely 1984). Early studies indicated the existence of fast and slow isoforms of all three subunits (Dabrowska et al. 1973; Perry 1974; Syska et al. 1974; Clarke et al. 1976; Dhoot and Perry 1979, 1980; Dhoot et al. 1978, 1979; Wilkinson and Grand 1978; Wilkinson 1980; Matsuda et al. 1981; Toyota and Shimada 1981).

Fast and slow TN-C isoforms of approximately 18 kDa (Collins 1974; Wilkinson 1980) are encoded by single copy genes in the human genome (Gahlmann et al. 1988). The slow TN-C isoform appears to be identical with the cardiac isoform (Wilkinson 1980; Gahlmann et al. 1988). The fast TN-C has two high-affinity, Ca^{2+}-specific binding sites (binding sites I and II) and two low-affinity Ca^{2+}/Mg^{2+} binding sites (binding sites III and IV). The slow TN-C displays alterations in the primary structure (Wilkinson 1980), leading to a strongly reduced Ca^{2+} affinity of binding site I (Potter and Johnson 1982). Two-dimensional electrophoresis reveals the presence of two charge variants for both the slow and fast TN-C isoforms (Fig. 9). These charge variants do not seem to result from different phosphorylation states because their apparent pIs are unaltered by alkaline phosphatase treatment (Härtner and Pette 1990).

Differences in the Ca^{2+} sensitivity between fast- and slow-twitch fibers (Moss 1982; Laszewski-Williams et al. 1989; Ruff 1989) may be related to the TN-C isoform content. Indeed, measurements on pCa/tension relations in permeabilised single rabbit muscle fibers demonstrate the dependence of tension development on specific TN-C isoforms (Moss et al. 1986). However, as discussed below, these differences may also relate partially to interactions with specific TN-T isoforms.

Slow and fast isoforms of TN-I are present as two molecular weight variants in slow- and fast-twitch muscles (Fig. 9). Only the fast TN-I isoform appears to be phosphorylatable as it exists in three charge variants which

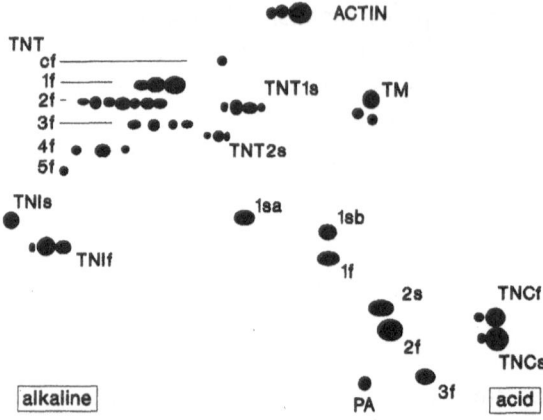

Fig. 9. Schematic illustration of two-dimensional electrophoretic separation of fast (*f*) and slow (*s*) isoforms of the three troponin subunits, troponin T (*TNT*), troponin I (*TNI*), and troponin C (*TNC*) in rabbit skeletal muscle. The diagram combines results obtained from two-dimensional nonequilibrium pH gradient gel electrophoresis (NEPHGE) for the separation of basic proteins and from two-dimensional gel electrophoresis for the separation of acidic proteins. *1f, 2f, 3f*, fast myosin light chains; *1sa, 1sb, 2s*, slow myosin light chains; *PA*, parvalbumin; *TM*, tropomyosin. (Modified from Härtner and Pette 1990)

are reduced to a single form after digestion with alkaline phosphatase (Härtner and Pette 1990).

The expression of a great number of fast TN-T isoforms is the result of alternative splicing of the TN-T primary RNA transcript (Wilkinson et al. 1984; Medford et al. 1984; Breitbart et al. 1985; Smillie et al. 1988). The fast TN-T gene of the rat contains 18 exons. Different splicing products result from five combinatorial exons (exons 4–8), two mutually exclusive exons (exons 16 and 17), and the remaining 11 constitutive exons. Combinatorial splicing mechanisms could theoretically generate up to 64 different mRNA isoforms (Breitbart et al. 1985; Breitbart and Nadal-Ginard 1986, 1987). The differential splicing of alternative and constitutive exons is developmentally controlled by muscle-specific trans-acting factors (Breitbart and Nadal-Ginard 1987).

The number of fast TN-T isoforms identified to date at the protein level is less than the number theoretically possible. Studies based upon differences in molecular weight show that the TN-T subunit exists in rabbit muscles in at least six fast and two slow isoforms (Briggs et al. 1987; Härtner et al. 1989). The identified fast TN-T isoforms in mammalian muscle consist of four major forms, i.e., TN-T$_{1f}$, TN-T$_{2f}$, TN-T$_{3f}$, and TN-T$_{4f}$, and two minor forms, i.e., TN-T$_{5f}$ and TN-T$_{cf}$ (Fig. 9). Analyses of the N-terminal amino acid sequence of the fast TN-T isoforms were performed by Briggs and Schachat (1989, see also Schachat et al. 1990). Their results

Table 4. Molecular weight variants and charge variants of fast and slow troponin subunit isoforms in rabbit hind limb muscles (Härtner and Pette 1990)

Troponin subunit	Molecular mass variants	Charge variants
$Tn-T_f$	6	19
$Tn-T_s$	2	7
$Tn-I_f$	1	3
$Tn-I_s$	1	1
$Tn-C_f$	1	2
$Tn-C_s$	1	2

indicated that $TN-T_{1f}$ is coded for by a mRNA containing transcripts of all alternatively spliced exons, whereas the $TN-T_{2f}$ mRNA is devoid of exon 4. The mRNA for $TN-T_{3f}$ contains neither exon 6 nor exon 7. Circumstantial evidence suggests that $TN-T_{4f}$ results from a combination of the splicing pathways leading to $TN-T_{2f}$ and $TN-T_{3f}$ (Briggs and Schachat 1989).

The slow TN-T isoforms are $TN-T_{1s}$ and $TN-T_{2s}$. The $TN-T_{1s}$ isoform has recently been identified as the "type I protein" (Schmitt and Pette 1988) which has previously been proposed as a specific marker for type I fibers (Heizmann et al. 1983). Additional charge variants (Fig. 9, Table 4) further increase the number of fast and slow TN-T isoforms. Most of the TN-T charge variants represent phosphorylated forms (Briggs et al. 1987; Härtner et al. 1989; Schachat et al. 1990).

The distribution of the various fast TN-T isoforms appears to follow a restricted pattern in association with the tropomyosin subunits (Moore and Schachat 1985; Moore et al. 1987), but perhaps not within specific fast fiber types (Schachat et al. 1985a; Briggs et al. 1987). The $TN-T_{1f}$ isoform is found in combination with the α/β-TM heterodimer, $TN-T_{2f}$ is preferentially coexpressed with the α-TM homodimer, and $TN-T_{3f}$ is found in combination with the α/β-TM heterodimer and the β-TM homodimer (Moore and Schachat 1985; Moore et al. 1987). The $TN-T_f$/TM combinations appear to be expressed in a continuum and, according to Schachat et al. (1985a, 1990), do not correlate with individual fast fiber subtypes in rabbit muscles. Single fiber analyses show that FG fibers of erector spinae muscle and the plantaris muscle predominantly express $TN-T_{2f}$, whereas $TN-T_{1f}$ is the major isoform in FG fibers of the diaphragm. In addition, FOG fibers of different muscles express $TN-T_{1f}$, $TN-T_{2f}$, $TN-T_{3f}$, and $TN-T_{4f}$ in highly variable proportions. Slow-twitch fibers contain almost exclusively slow TN-T isoforms ($TN-T_{1s}$, $TN-T_{2s}$). Small amounts of slow TN-T isoforms are also found in fast-twitch fibers (Moore et al. 1987).

   These observations might indicate the existence of muscle-specific rather than fiber type-specific distribution patterns. This suggestion is corroborated by recent observations on the TN-T isoform distribution in extraocular muscles of the rabbit. Although TN-T$_{3f}$ and TN-T$_{2s}$ are minor isoforms in limb and trunk musculature, singly innervated extraocular muscle fibers predominantly contain the TN-T$_{3f}$ isoform, whereas TN-T$_{2s}$ predominates in multiply innervated fibers which exert tonic contractions (Briggs et al. 1988). Muscle-specific distribution patterns of fast TN-T isoforms have also been shown in fast avian muscles. Five fast TN-T isoforms (L1−L5) varying in molecular weight are present in fast-twitch leg muscles of adult chicken, whereas adult pectoral muscle contains, in addition to L3, L4, and L5, the heavier fast TN-T isoforms B1 and B2 (Dhoot 1988).

   Many questions still remain about the distribution patterns of the TN-T isoforms. The elaborate results of Schachat and coworkers have been derived from a fiber classification scheme which is primarily based upon metabolic properties and which identifies FG, FOG, and SO fiber types (Moore and Schachat 1985; Schachat et al. 1985a; Moore et al. 1987). Their attempt to correlate the distribution of the fast TN-T isoforms with fiber types may have led to ambiguous results, and a different picture might emerge with a myosin-based fiber classification scheme (Schmitt and Pette 1990). Thus, fast and slow TN-T isoforms are correlated with fast and slow myosin heavy chain isoforms in rabbit muscles. Moreover, in the fast-twitch fiber population myosin HCIIb is always coexpressed with TN-T$_{2f}$ and HCIIa with TN-T$_{3f}$. There is evidence that HCIId coexists with TN-T$_{3f}$ and TN-T$_{4f}$. Type I fibers containing the slow HCI isoform display the slow TN-T$_{1s}$ together with minor amounts of TN-T$_{2s}$ (Schmitt and Pette 1990).

   A functional significance has been assigned on the basis of the pCa/tension relationship to the different fast TN-T isoforms in combination with the α- and β-TM subunits (Schachat et al. 1987, 1990; Greaser et al. 1988). Fibers containing predominantly the TN-T$_{2f}$ isoform and the α_2-TM homodimer exhibit steeper pCa/tension relations than fibers containing predominantly either the TN-T$_{1f}$ or TN-T$_{3f}$ isoforms in combination with the α/βTM heterodimer (Fig. 10). A relationship between Ca^{2+} sensitivity and the expression of specific TN-T isoforms has also been observed in chicken muscle (Reiser et al. 1987b).

   Insights into the functional significance of TN-T isoforms have also arisen from time course studies of rabbit muscle fast-to-slow transformation induced by chronic low-frequency stimulation. Successive reductions occur in the fast TN-T isoforms (Schachat et al. 1988; Härtner et al. 1989) with a progressive rise in the slow TN-T isoforms (Härtner et al. 1989; Pette

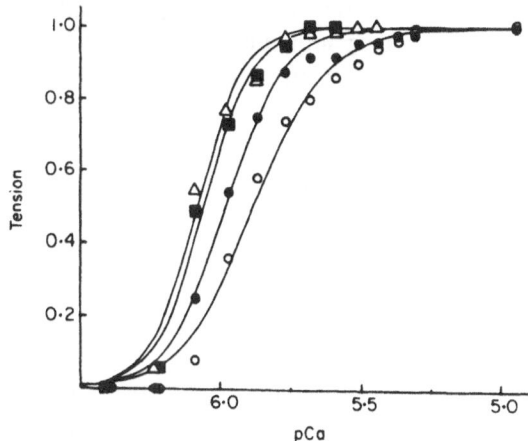

Fig. 10. pCa/tension relations of skinned single fibers from various rabbit muscles with different troponin–tropomyosin complements. The plantaris (*open triangles*) and psoas (*filled squares*) fibers, expressing predominantly TN-T$_{2f}$ (85% – 89%) and α_2-TM, have half-maximal pCa values of 6.08 and 6.06, respectively. Both α-TM and TN-T$_{2f}$ amount to approximately 50% of total tropomyosin and TN-T species, respectively, in the extensor digitorum longus fiber (*filled circles*), which exhibits a half-maximal pCa value of 5.98. The sartorius fiber (*open circles*), with only 38% of α-TM, contains TN-T$_{3f}$ as the major (53%) TN-T species and has a half-maximal pCa value of 5.88. (From Schachat et al. 1987)

1990). The ordered sequence of the TN-T isoform exchanges (sequential decreases in TN-T$_{2f}$, TN-T$_{4f}$, TN-T$_{1f}$, and TN-T$_{3f}$) with concomitant increases in TN-T$_{1s}$ and TN-T$_{2s}$ corresponds to the time pattern of stimulation-induced fiber type transformations (Fig. 7). Type IIB fibers decrease first and this coincides with the disappearance of the TN-T$_{2f}$ isoform. The intermediate and transient increases in fiber types IIA, IIC, and IC suggest that these fibers mainly contain, in addition to slow TN-T isoforms, the fast isoforms TN-T$_{1f}$ and TN-T$_{3f}$. A progressive, although incomplete, exchange of the fast with the slow α-TM subunit occurs in long-term stimulated rabbit fast-twitch muscle (Härtner et al. 1989). The increase in the slow α-TM subunit appears to occur concomitantly with the increase in the slow isoforms TN-T$_{1s}$ and TN-T$_{2s}$. However, the fast-to-slow transition of the α-TM subunit is less extensive than that of the TN-T isoforms. Like normal soleus muscles, long-term stimulated fast-twitch muscles contain the slow and fast α-TM subunits together with TN-T$_{1s}$ and TN-T$_{2s}$ but only traces of the fast TN-T isoforms are present (Härtner et al. 1989). This observation suggests that slow TN-T isoforms may be capable of interacting not only with the slow α-TM subunit isoform, but also with the fast isoform. Therefore, a whole spectrum of variable combinations between fast and slow TM and TN-T isoforms may exist in transforming fibers.

The changes occurring during the induced fast-to-slow transformation in the other troponin subunits, TN-C and TN-I, are less pronounced than those of TN-T (Härtner and Pette 1988, 1990; Pette 1990). As such, transforming fibers may contain troponin molecules made up of both fast and slow subunits. Fast and slow isoforms of TN-T, as well as TN-I and TN-C, are also coexpressed in histochemically defined C fibers in rabbit (Salviati et al. 1982) and human muscles (Salviati et al. 1983; Schantz and Dhoot 1987).

3.2.5 α-Actinin

Different fiber types have been shown ultrastructurally to contain different Z disc arrangements and widths (Gauthier 1969; Rowe 1973; Eisenberg 1974; Gauthier 1974; Eisenberg and Kuda 1976; Tomanek 1976; Hikida 1978; Eisenberg and Salmons 1981; Sjöström et al. 1982a, b; Eisenberg 1983; Hoppeler 1986; Thornell et al. 1987). As judged from differences in aerobic oxidative potential, low-oxidative fibers have thin Z discs, whereas oxidative fibers have thick Z discs. The major component of the Z disc is α-actinin, an actin-anchoring protein (Ebashi and Ebashi 1965; Maruyama and Ebashi 1965; Masaki et al. 1967; Takahashi and Hattori 1989), a dimer with subunits of approximately 100 kDa (Langner and Pepe 1980; Feramisco and Burridge 1980). The actin filament-binding domain of α-actinin is highly conserved between slime molds and man (Noegel et al. 1987; Schleicher et al. 1988). It exhibits homologies to spectrin and calmodulin (Baron et al. 1987) and contains spectrin-like repeats as dystrophin (Hoffman et al. 1987; Hammonds 1987; Davison and Critchley 1988; Koenig et al. 1988; Schleicher et al. 1988).

Three isoforms, all homodimers, have been described and appear to be muscle-specific (Kobayashi et al. 1984), two fast α-actinins (types I and II) and one slow (Kobayashi et al. 1983). The slow α-actinin differs immunologically from one of the fast actinins (Kobayashi et al. 1983). More recent evidence suggests that the slow isoform is identical with one of the fast isoforms, i.e., α-actinin$_{1f/s}$ (Schachat et al. 1985b). The other fast isoform, α-actinin$_{2f}$, has recently been shown to react with antisera directed against dystrophin (Hoffman et al. 1989).

The distribution of the two α-actinin isoforms (α-actinin$_{1f/s}$, α-actinin$_{2f}$) does not correlate with fast fiber types histochemically defined as FG or FOG (Schachat et al. 1985b). The α-actinin$_{1f/s}$ isoform is predominant in most FG and FOG fibers, while α-actinin$_{2f}$ predominates in a few specific FG fibers, e.g., FG fibers from rabbit psoas muscle. Indeed, Schachat et al. (1985b) found that both α-actinins (α-actinin$_{1f/s}$ and α-actinin$_{2f}$) can be coexpressed in the same fiber. However, the α-actinin

Fig. 11. Correlation between the α-actinin$_{1f/s}$ isoform content and fast troponin-T (*Th T*) isoform distribution in various fast-twitch muscles of the rabbit. Each point represents a determination of myofibrils from a single muscle. *Ad*, adductor magnus; *EDL*, extensor digitorum longus. (From Schachat et al. 1985b)

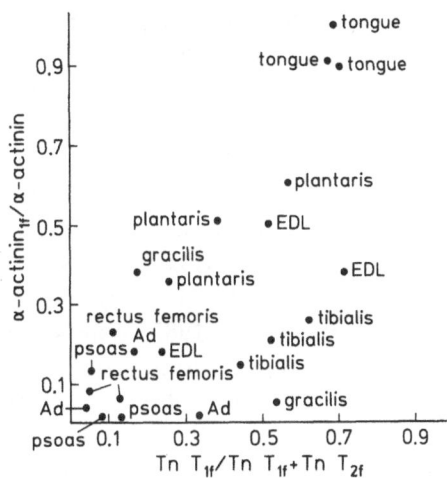

isoforms appear to be expressed in a specific manner with specific troponin/tropomyosin complexes and Z disc widths (Schachat et al. 1985b). Thus, fibers expressing α-actinin$_{1f/s}$ exhibit thick Z discs, whereas fibers expressing α-actinin$_{2f}$ are characterized by thin Z discs.

A relationship also appears to exist between α-actinin and fast TN-T isoform expression (Schachat et al. 1985a). The TN-T$_{1f}$ and TN-T$_{3f}$ isoforms tend to be coexpressed with the α-actinin$_{1f/s}$ isoform, whereas TN-T$_{2f}$ tends to be combined with α-actinin$_{2f}$ (Fig. 11). This relationship is maintained in rabbit muscle undergoing a fast-to-slow transformation (Schachat et al. 1988, 1990).

3.2.6 Other Myofibrillar Proteins

Limited information exists on isoforms of other myofibrillar proteins. Several muscle proteins (M-protein, C-protein, H-protein, and X-protein), located at specific positions on the thick filament, appear to have muscle- and fiber-specific isoforms. No such information exists for other myofibrillar proteins, e.g., β-actinin (Asami et al. 1988a, b; Funatsu et al. 1988) and various scaffold proteins (see Ohtsuki et al. 1986).

Three different proteins constitute the M band, the 165-kDa M protein (Eaton and Pepe 1972; Masaki and Takaiti 1972, 1974), the 185-kDa myomesin (Eppenberger et al. 1981; Grove et al. 1985), and MM-creatine kinase (Turner et al. 1973; Wallimann et al. 1983). Myomesin seems to exist in different isoforms in fast and slow chicken muscle fibers (Grove et al. 1989). Specific M band patterns have been observed ultrastructurally for different fiber types (Sjöström et al. 1982b; Wallimann and Eppenberger

1985; Thornell et al. 1987, 1990) and may be related to the different M-protein isoforms detected electrophoretically in single fibers (Salviati et al. 1982; Staron and Pette 1987a). In addition, immunohistochemical evidence from rat soleus muscle spindles indicates that each intrafusal fiber type has a unique composition of M band proteins (Pedrosa et al. 1989).

The 140-kDa C-protein is a component of the thick filament (Starr and Offer 1971; Offer et al. 1973; Pepe and Drucker 1975; Craig and Offer 1976). Various isoforms of C-protein have been elucidated for skeletal (two isoforms, i.e., white and red) and cardiac muscles (Callaway and Bechtel 1981; Reinach et al. 1982; Yamamoto and Moos 1983; Takano-Ohmuro et al. 1989) and appear to be sequentially expressed during chicken pectoralis muscle development (Obinata et al. 1984). The immunohistochemically assessed distribution of the two skeletal muscle isoforms does not strictly follow the mATPase-based histochemical fiber typing (Dhoot et al. 1985). Both skeletal muscle isoforms may coexist in type II rabbit fibers (Dhoot et al. 1985) and may even be coexpressed within a single sarcomere in fast-twitch chicken muscle fibers (Reinach et al. 1983; Dennis et al. 1984). According to Locker and Wild (1986), three C-protein isoforms (white, α-red, and β-red) exist in skeletal muscle. The α-red C-protein may be identical with the X-protein (Locker and Wild 1986). More recently at least six isoforms of C-protein have been elucidated in chicken muscle: four adult and two embryonic isoforms (Takano-Ohmuro et al. 1989). These consist of fast-type isoform (CF) from pectoralis muscle, two slow-type isoforms (one from the slow-tonic anterior latissimus dorsi muscle, CS3; one from the posterior latissimus dorsi muscle, CS4, a cardiac-type (Cc), and two slow-type isoforms (CS1 and CS2) expressed during development of the pectoralis muscle.

Two other myofibrillar proteins, a 74-kDa H-protein (Starr and Offer 1983; Yamamoto 1984) and a 135-kDa X-protein (Starr et al. 1980), have been isolated from rabbit skeletal muscle and localized as components of the thick filament (Starr et al. 1985). The distribution of C-, H-, and X-proteins is such that fast white fibers contain C- and H-proteins, but little or no X-protein. Conversely, slow fibers contain X-protein, but very little C- or H-proteins. Fast red fibers usually contain all three proteins in varying amounts (Starr and Offer 1983; Starr et al. 1985).

Another myofibrillar protein has recently been isolated from rabbit skeletal muscle and localized in the Z disc of all type I fibers and some type II fibers (Chen et al. 1986). Although the function of this 32-kDa protein is unknown, it remains to be demonstrated whether or not this protein is identical with carbonic anhydrase III, a protein found to be diffusely distributed in the cytoplasm (Väänänen et al. 1986; Frémont et al. 1988). As previously discussed, carbonic anhydrase III correlates with the proportion of type I fibers (Väänänen et al. 1982; Gagnon et al. 1985; Jeffery et al. 1986, 1987, 1990; Frémont et al. 1987).

Scaffold proteins are part of the cytoskeletal system of muscle fibers and consist of transverse and longitudinal filaments located outside and inside of the sarcomere (Wang and Ramirez-Mitchell 1983; Pierobon-Bormioli et al. 1989). There is a paucity of information on these cytoskeletal proteins. However, recent evidence suggests that the giant myofibrillar proteins of the endosarcomeric system exist in size variants which might represent tissue-specific isoforms (Hill and Weber 1986; Wang and Wright 1988; Akster et al. 1989). These high molecular weight proteins named titin (Wang et al. 1979) or connectin (Maruyama et al. 1977, 1989) and nebulin (Wang and Williamson 1980) have been implicated in passive myofibrillar elasticity (Locker and Leet 1976; Wang 1984), in controlling thick filament position during contraction (Horowits et al. 1986; Horowits and Podolsky 1987), and as templates for the assembly and regulation of the length of thin and thick filaments (Wang and Wright 1988).

3.3 Ca^{2+}-Sequestering Proteins

Calcium is the crucial link between excitation and contraction (see de Meis 1981; Martonosi and Beeler 1984; Rüegg 1988). Its intracellular concentration is finely controlled in skeletal muscle by release from the terminal cisternae of the sarcoplasmic reticulum (SR) and by the Ca^{2+}-sequestering function of the longitudinal SR. The specialization of the Ca^{2+}-handling system in different mammalian fiber types is clearly illustrated by the differences in Ca^{2+} transients detected by microinjecting aequorin into fast- and slow-twitch rat muscle fibers (Eusebi et al. 1980).

Numerous studies have demonstrated pronounced differences in Ca^{2+} uptake by the SR between fast- and slow-twitch muscles (Sréter and Gergely 1964; Harigaya et al. 1968; Sréter 1969; Fiehn and Peter 1971; Margreth et al. 1972, 1974; Salmons and Sréter 1976; Briggs et al. 1977; Heilmann et al. 1977; Mabuchi and Sréter 1978; van Winkle and Schwartz 1978; van Winkle et al. 1978; Heilmann and Pette 1979; Kim et al. 1981; Zubrzycka-Gaarn et al. 1982; Leberer et al. 1988). These differences, which have also been shown for single human muscle fibers (Salviati et al. 1983), partially relate to variations in the membrane surface area of the SR and in the content (per membrane area) of the Ca^{2+}-transport ATPase in different fiber types. Morphometric analyses of muscles from several species (Luff and Atwood 1971; Stonnington and Engel 1973; Tomanek et al. 1973; van Winkle and Schwartz 1978; Schmalbruch 1979; Davey and Wong 1980; Eisenberg 1983; Ogata and Yamasaki 1985; Eisenberg et al. 1987) indicate that type I fibers contain the lowest and type II fibers the highest volume percentage of SR. In addition, distinct differences exist between specific muscle fibers in guinea pig muscles in the surface area of terminal cisternae per cell volume (Eisenberg et al. 1974; Eisenberg and Kuda 1975). Although pronounced overlap exists, the FG fibers have the highest mean values, the FOG fibers have intermediate values, and the SO fibers have the lowest values. In rabbit muscle, a negative correlation exists between the size of the T tubule system per fiber volume and the Z disc width (Eisenberg and Salmons 1981). Furthermore, the fast-twitch fibers in guinea pig muscles have a much higher proportion of junctional segments in the T tubule network than slow-twitch fibers. These differences account for a twofold higher density of junctional "feet" in fast-twitch fibers than in slow-twitch fibers (Franzini-Armstrong et al. 1988).

The relationship between contractile speed and the development of the major Ca^{2+}-regulatory membrane system in a muscle fiber adds to the concept of muscle-specific fiber types. Comparative analyses of slow-twitch fibers in different cat muscle indicate variations in the fractional volume of the SR (Schmalbruch 1979). In addition, a combined morphometric and physiological study on rat tibialis anterior and soleus muscles has shown an inverse relationship between the volume density of the terminal cisternae and isometric contraction time over a wide range of motor units (Kugelberg and Thornell 1983).

3.3.1 Ca^{2+}-Transport ATPase of the Sarcoplasmic Reticulum

The sequestration of Ca^{2+} from the sarcoplasm back into the sarcoplasmic reticulum is an energy-dependent process and is accomplished by the 115-kDa Ca^{2+}/Mg^{2+}-dependent transport ATPase (Ca^{2+}-ATPase, Hasselbach 1964), predominantly localized in the longitudinal SR (Jorgensen et al. 1982; Jorgensen and McGuffee 1987). Early studies on preparations of fragmented SR from muscles with a predominant fast or slow fiber type distribution have demonstrated markedly lower Ca^{2+}-ATPase activities in slow- twitch muscles than in fast-twitch muscles (Sréter 1969; Margreth et al. 1972; Ramiirez and Pette 1974; Heilmann et al. 1977, 1981; van Winkle et al. 1978; Heilmann and Pette 1979; Zubrzycka-Gaarn et al. 1982). These results are partially supported by ultrastructural freeze fracture analysis demonstrating differences in the density of 7−9 nm intramembranous particles per unit area in SR preparations from fast- and slow-twitch muscles (Beringer 1975; Bray and Rayns 1976; Ryan and Shafiq 1980; Heilmann et al. 1981; Zubrzycka-Gaarn et al. 1982; Franzini-Armstrong 1984; Gauthier and Hobbs 1986; Ferguson and Franzini-Armstrong 1988). However, the difference in the number of intramembranous particles per unit area (two- to threefold) between the SR from fast- and slow-twitch fibers appears to be less pronounced when compared with the differences in Ca^{2+} uptake, Ca^{2+}-ATPase activity, or the Ca^{2+}-ATPase content of whole muscle.

Electron microscopic analyses of crystalline arrays of Ca^{2+}-ATPase vanadate complexes reveal that the low Ca^{2+}-transport activity of slow-twitch muscles is due to the presence of large amounts of non-SR membrane elements (Dux and Martonosi 1984). Although the Ca^{2+}-ATPase-containing membranes show similar concentrations of the enzyme per membrane area in both slow- and fast-twitch muscle, the fraction of the Ca^{2+}-ATPase-containing membranes is much lower in microsomal preparations from slow-twitch muscles than from fast-twitch muscles (Dux and Martonosi 1984). Thus, it appears that the Ca^{2+}-sequestering intracellular membrane system is confined to a smaller membrane fraction in slow-twitch fibers.

This difference in the distribution of Ca^{2+}-ATPase between fast- and slow-twitch muscles is also confirmed by immunochemical quantification of the Ca^{2+}-ATPase protein in isolated SR preparations (Dulhunty et al. 1987) and in homogenates from muscles of defined fiber composition (Leberer and Pette 1986a). Unlike measurements on isolated SR preparations, measurements on whole muscle homogenates avoid errors due to the inability to completely isolate SR and also the difficulty in obtaining pure SR membranes. Moreover, this method makes it possible to assess SR Ca^{2+}-ATPase tissue levels and, thus, compare the major functional element of the SR at the cellular level. Rabbit muscles composed predominantly of type II fibers display six to seven times more Ca^{2+}-ATPase protein than slow-twitch muscles. Furthermore, a slightly higher Ca^{2+}-ATPase concentration is found in fast-twitch muscles containing a higher percentage of IIB fibers than IIA fibers (Leberer and Pette 1986a). This difference in the Ca^{2+}-ATPase distribution between fast- and slow-twitch muscles is less pronounced in the mouse. Adult fast-twitch muscles of the mouse hindlimb contain approximately three times more Ca^{2+}-ATPase than soleus muscle (Leberer et al. 1988). This difference is smaller than in the rabbit because mouse soleus muscle contains a high percentage (up to 50%) of fast-twitch fibers.

Antibody studies indicate that differences exist between the Ca^{2+}-ATPase of different muscles, especially fast- and slow-twitch muscles (DeFoor et al. 1980; Damiani et al. 1981; Volpe et al. 1982; Leberer and Pette 1986a; Kaprielian and Fambrough 1987). Indeed, recombinant DNA technology has revealed that two separate genes encode for fast-twitch and slow-twitch/cardiac isoforms of the SR Ca^{2+}-ATPase (Brandl et al. 1986, 1987; Korczak et al. 1988). However, the diversity of the two genes is restricted to only 15% of the total sequence (Brandl et al. 1986). This extensive homology may explain the difficulty in raising antibodies that specifically react with either isoform.

Immunohistochemical studies give results similar to those obtained by immunochemical quantification of the Ca^{2+}-ATPase in whole muscle homogenates (Leberer and Pette 1986a). Light and electron microscopic investigations indicate that type II fibers have a markedly higher SR Ca^{2+}-ATPase content than type I fibers (Jorgensen et al. 1979, 1982, 1983; Maier et al. 1986b; Dulhunty et al. 1987; Krenács et al. 1989). Type IIB fibers consistently display a slightly higher content of Ca^{2+}-ATPase than type IIA fibers in both rabbit and rat muscles (Maier et al. 1986b; Krenács et al. 1989). Moreover, some heterogeneity appears to exist within the type I fiber populations of rabbit and rat soleus muscles (Maier et al. 1986b).

Ca^{2+} uptake by the SR in cardiac muscle is under the regulatory influence of phospholamban, a phosphorylatable regulatory protein of Ca^{2+}-ATPase (Tada et al. 1975; Tada and Inui 1983). Phospholamban is also expressed in slow-twitch, but not in fast-twitch muscles (Kirchberger and Tada 1976; Heilmann et al. 1977; Jorgensen and Jones 1986; Leberer et al. 1989). The nucleotide sequences of cDNA clones encoding phospholamban seem to be virtually identical for clones isolated from rabbit slow-twitch and cardiac muscle libraries. Therefore, both types of muscle express the same phospholamban gene (Fuji et al. 1988), and the possibility exists that phospholamban also plays a regulatory role in Ca^{2+} sequestration in slow-twitch fibers.

3.3.2 Calsequestrin

Ca^{2+} ions, pumped by Ca^{2+}-ATPase into the lumen of the SR, are bound to calsequestrin (MacLennan and Wong 1971) in the terminal cisternae (Jorgensen et al. 1983, 1985; MacLennan et al. 1983). Calsequestrin is expressed in fast- and slow-twitch muscles and in cardiac muscles. Isoforms of this major Ca^{2+}-binding protein have been distinguished between skeletal and cardiac muscles (Fliegel et al. 1987). Although these two isoforms are products of different genes, they are 65% identical (Scott et al. 1988). In the rabbit, the predominant calsequestrin isoform in slow-twitch muscle appears to be identical with the isoform in fast-twitch muscle (Damiani et al. 1986; Fliegel et al. 1989).

Immunochemical measurements of calsequestrin content in homogenates from rabbit (Leberer and Pette 1986a; Leberer et al. 1986) and mouse (Leberer et al. 1988) muscles with defined fiber composition reveal approximately twice as much calsequestrin in type II fibers as in type I. No differences are detectable in the calsequestrin content of type II A and type II B fibers of rabbit muscles. The use of immunohistochemistry gives similar results in both rat and rabbit muscles, i.e., higher calsequestrin content in fast-twitch fibers than in slow-twitch fibers, but no difference between fast fiber subtypes (Maier et al. 1986b).

3.3.3 Parvalbumin

Parvalbumin is an acidic Ca^{2+}- and Mg^{2+}-binding protein of 12 kDa. In skeletal muscle, it is thought to play a role in muscle relaxation and to act as a cytosolic Ca^{2+} buffer. Parvalbumin appears to be functionally inter-calated between troponin C and the Ca^{2+}-transport ATPase of the SR (Gillis et al. 1982; Heizmann 1984; Gillis 1985; Klug et al. 1988). In various mammalian muscles, parvalbumin is distributed almost exclusively in fast-twitch muscles (Blum et al. 1977; Heizmann et al. 1982; Klug et al. 1983a, b; Berchtold et al. 1984; Leberer and Pette 1986a, b, 1990; Müntener et al. 1987; Berchtold 1989). In addition, its concentration is higher in fast-twitch muscles of small mammals than in fast-twitch muscles of large mammals (Heizmann et al. 1982; Simoneau et al. 1989). It appears that its function as an intermediate Ca^{2+}-trapping protein is more important in the faster relaxing muscles of small mammals, where Ca^{2+}-transport ATPase may be a limiting factor in the relaxation process. This suggestion is supported by the observation that the duration of the active state is prolonged under experimental conditions with reduced parvalbumin content (Klug et al. 1988; Pette 1989; Simoneau et al. 1989). Immunohistochemistry of rat muscles reveals high parvalbumin levels in type II B fibers, variable and low levels in type II A fibers, and extremely low levels in type I fibers (Celio and Heizmann 1982). These results are in agreement with the immunochemical quantitation of parvalbumin in homogenates from rat and rabbit muscles with defined fiber composition (Leberer and Pette 1986a, 1990; Leberer et al. 1986, 1987; Gundersen et al. 1988). In rabbit and rat, parvalbumin contents are 200- to 300-fold higher in fast-twitch muscles than in slow-twitch muscles. Significantly higher parvalbumin contents are found in fast-twitch muscles with a higher percentage of type II B fibers than in fast-twitch muscles with a higher percentage of type II A fibers.

4 Conclusions

Muscle fibers are not static structures, but represent versatile entities adapting to altered functional demands, hormonal signals, and changes in neural input. Their dynamic nature makes it difficult to categorize them into distinct units. Therefore, applying any type of rigid structural and/or functional confines results in an oversimplification which does not take into account the plasticity of phenotypic expression. Still, fiber typing is an extremely useful tool in many fields of muscle biology. Classification schemes have helped to define functional, metabolic, and molecular properties of muscle fibers selected on the basis of their histochemical appearance. Under such conditions, certain muscle fiber populations with similar functional properties and molecular profiles may be delineated, e.g., fast- and slow-twitch fibers.

It is obvious that more than two groups can be distinguished, especially with the use of additional qualitative and quantitative analytical methods. However, it must be realized that some molecular and/or functional properties may change without affecting others or without changing the histochemical appearance of a given fiber (muscle specificity, species specificity, developmental or adaptive processes, pathological conditions) and that innumerable fiber type transients exist. Therefore, a distinction between specific groups or types, although possible, must strictly refer to the method upon which it is based. Moreover, the complexity and dynamic nature of muscle makes it imperative that future investigators refrain from drawing conclusions about single muscle fibers when using broad classification schemes (e.g., "white" and "red").

Major fiber types represent large groups of muscle fibers displaying similar phenotypes. Such groups may reflect stable programs or stages of gene expression with preferential combinations of sarcoplasmic and myofibrillar protein isoforms. Moreover, they may be the direct result of predominant patterns of functional demand and/or neuromuscular activity. As such, fibers within a particular group may contain similar structural and functional properties, but the most similar muscle fibers will be those within the same motor unit. Theoretically, there will be at least as many fiber types as there are motor units in a muscle. Collectively, within a muscle a fiber type may exhibit a continuum of structural and functional properties which overlap with other fiber types. The multiplicity of fiber phenotypes is increased by the possibility of local factors modulating gene expression within motor units and along the length of single fibers. As increased numbers of different muscles and species are investigated and more refined high-resolution techniques are developed and applied, the full extent of cellular and molecular diversity of skeletal muscle should become evident.

References

Akster HA, Granzier HL, Focant B (1989) Differences in I band structure, sarcomere extensibility, and electrophoresis of titin between two muscle fiber types of the perch (*Perca fluviatilis l.*). J Ultrastructure Molecul Structure Res 102:109–121

Alway SE, MacDougall JD, Sale DG, Sutton JR, McComas AJ (1988) Functional and structural adaptations in skeletal muscle of trained athletes. J Appl Physiol 64:1114–1120

Andersen P, Henriksson J (1977) Training induced changes in the subgroups of human type II skeletal muscle fibers. Acta Physiol Scand 99:123–125

Andreadis A, Gallego ME, Nadal-Ginard B (1987) Generation of protein isoform diversity by alternative splicing: mechanistic and biological implications. Ann Rev Cell Biol 3:207–242

Arndt I, Pepe FA (1975) Antigenic specificity of red and white muscle myosin. J Histochem Cytochem 23:159–168

Asami Y, Funatsu T, Ishiwata S (1988a) Transition of β-actinin isoforms during development of chicken skeletal muscle. J Biochem (Tokyo) 103:72–75

Asami Y, Funatsu T, Ishiwata S (1988b) β-Actinin isoforms in various types of muscle and non-muscle tissues. J Biochem (Tokyo) 103:76–80

Askanas V, Engel WK (1975) Distinct sub-types of type I fibres of human skeletal muscle. Neurology (Minneapolis) 25:879–887

Bader D, Masaki T, Fischman DA (1982) Immunochemical analysis of myosin heavy chain during myogenesis in vivo and in vitro. J Cell Biol 95:763–770

Baldwin KM, Klinkerfuss GH, Terjung RL, Molé PA, Holloszy JO (1972) Respiratory capacity of white, red, and intermediate muscle: adaptive response to exercise. Am J Physiol 222:373–378

Balint M, Sréter FA, Gergely J (1975) Fragmentation of myosin by papain – studies on myosin from adult fast and slow skeletal and cardiac, and embryonic muscle. Arch Biochem Biophys 168:557–566

Bandman E (1985a) Continued expression of neonatal myosin heavy chain in adult dystrophic skeletal muscle. Science 227:780–782

Bandman E (1985b) Myosin isoenzyme transitions in muscle development, maturation, and disease. Int Rev Cytol 97:97–131

Bandman E, Matsuda R, Strohman RC (1982a) Myosin heavy chains from two different adult fast-twitch muscles have different peptide maps but identical mRNAs. Cell 29:645–650

Bandman E, Matsuda R, Strohman RC (1982b) Developmental appearance of myosin heavy and light chain isoforms in vivo and in vitro in chicken skeletal muscle. Dev Biol 93:508–518

Bär A, Pette D (1988) Three fast myosin heavy chains in adult rat skeletal muscle. FEBS Lett 235:153–155

Bárány M (1967) ATPase activity of myosin correlated with speed of muscle shortening. J Gen Physiol 50:197–218

Bárány M, Bárány K, Reckard T, Volpe A (1965) Myosin of fast and slow muscles of the rabbit. Arch Biochem Biophys 109:185–191

Barnard RJ, Edgerton VR, Furukawa T, Peter JB (1971) Histochemical, biochemical and contractile properties of red, white, and intermediate fibers. Am J Physiol 220:410–414

Baron MD, Davison MD, Jones P, Critchley DR (1987) The sequence of chick α-actinin reveals homologies to spectrin and calmodulin. J Biol Chem 262:17623–17629

Barton PJR, Buckingham ME (1985) The myosin alkali light chain proteins and their genes. Biochem J 231:249–261

Barton PJR, Cohen A, Robert B, Fiszman MY, Bonhomme F, Guenet J-L, Leader DP, Buckingham ME (1985a) The myosin alkali light chains of mouse ventricular and slow skeletal muscle are indistinguishable and are encoded by the same gene. J Biol Chem 260: 8578–8584

Barton PJR, Robert B, Fiszman MY, Leader DP, Buckingham ME (1985b) The same myosin alkali light chain gene is expressed in adult cardiac atria and in fetal skeletal muscle. J Muscle Res Cell Motil 6:461–475

Bass A, Brdiczka D, Eyer P, Hofer S, Pette D (1969) Metabolic differentiation of distinct muscle types at the level of enzymatic organization. Eur J Biochem 10:198–206

Bass A, Lusch G, Pette D (1970) Postnatal differentiation of the enzyme activity pattern of energy-supplying metabolism in slow (red) and fast (white) muscles of chicken. Eur J Biochem 13:289–292

Bauman H, Cao K, Howald H (1984) Improved resolution with one-dimensional polyacryl-amide gel electrophoresis: myofibrillar proteins from typed single fibers of human muscle. Anal Biochem 137:517–522

Baumann H, Jaeggi M, Soland F, Howald H, Schaub MC (1987) Exercise training induces transitions of myosin isoform subunits within histochemically typed human muscle fibres. Pflügers Arch 409:349–360

Berchtold MW (1989) Structure and expression of genes encoding the three-domain Ca^{2+}-binding proteins parvalbumin and oncomodulin. Biochem Biophys Acta 1009: 201–215

Berchtold MW, Celio MR, Heizmann CW (1984) Parvalbumin in non-muscle tissues of the rat. Quantitation and immunohistochemical localization. J Biol Chem 259:5189–5196

Beringer T (1975) A freeze-fracture study of sarcoplasmic reticulum from fast and slow muscle of the mouse. Anat Rec 184:647–664

Billeter R, Weber H, Lutz H, Howald H, Eppenberger HM, Jenny E (1980) Myosin types in human skeletal muscle fibres. Histochemistry 65:249–259

Billeter R, Heizmann CW, Howald H, Jenny E (1981a) Analysis of myosin light and heavy chain types in single human skeletal muscle fibers. Eur J Biochem 116:389–395

Billeter R, Heizmann CW, Reist U, Howald H, Jenny E (1981b) α- and β-tropomyosins in typed single fibers of human skeletal muscle. FEBS Lett 132:133–136

Billeter R, Heizmann CW, Reist U, Howald H, Jenny E (1982) Two-dimensional peptide analyses of myosin heavy chains and actin from single-typed human skeletal muscle fibers. FEBS Lett 139:45–48

Biral D, Damiani E, Volpe P, Salviati G, Margreth A (1982) Polymorphism of myosin light chains – an electrophoretic and immunological study of rabbit skeletal-muscle myosins. Biochem J 203:529–540

Biral D, Damiani E, Margreth A, Scarpini E (1984) Myosin subunit composition in human developing muscle. Biochem J 224:923–931

Biral D, Betto R, Danieli-Betto D, Salviati G (1988) Myosin heavy chain composition of single fibres from normal human muscle. Biochem J 250:307–308

Biral D, Scarpini E, Angelini C, Salviati G, Margreth A (1989) Myosin heavy chain composition of muscle fibers in spinal muscular atrophy. Muscle Nerve 12:43–51

Blum HE, Lehky P, Kohler L, Stein EA, Fischer EH (1977) Comparative properties of vertebrate parvalbumins. J Biol Chem 252:2834–2838

Brandl CJ, Green MN, Korczak B, MacLennan DH (1986) Two Ca^{2+} ATPase genes: homologies and mechanistic implications of deduced amino acid sequences. Cell 44:597–607

Brandl CJ, deLeon S, Martin DR, MacLennan DH (1987) Adult forms of the Ca^{2+} ATPase of sarcoplasmic reticulum. Expression in developing skeletal muscle. J Biol Chem 262:3768–3774

Bray DF, Rayns DG (1976) A comparative freeze-etch study of the sarcoplasmic reticulum of avian fast and slow muscle fibres. J Ultrastruct Res 57:251–259

Breitbart RE, Nadal-Ginard B (1986) Complete nucleotide sequence of the fast skeletal troponin T gene. Alternatively spliced exons exhibit unusual interspecies divergence. J Mol Biol 188:313–324

Breitbart RE, Nadal-Ginard B (1987) Developmentally induced, muscle-specific trans factors control the differential splicing of alternative and constitutive troponin T exons. Cell 49:793–803

Breitbart RE, Nguyen HT, Medford RM, Destree AT, Mahdavi V, Nadal-Ginard B (1985) Intricate combinatorial patterns of exon splicing generate multiple regulated troponin T isoforms from a single gene. Cell 41:67–82

Briggs FN, Poland JL, Solaro RJ (1977) Relative capabilities of sarcoplasmic reticulum in fast and slow mammalian skeletal muscles. J Physiol (Lond) 266:587–594

Briggs MM, Schachat F (1989) N-terminal amino acid sequences of three functionally different troponin T isoforms from rabbit fast skeletal muscle. J Mol Biol 206:245–249

Briggs MM, Lin JJ-C, Schachat FH (1987) The extent of amino-terminal heterogeneity in rabbit fast skeletal muscle troponin T. J Muscle Res Cell Motil 8:1–12

Briggs MM, Jacoby J, Davidowitz J, Schachat FH (1988) Expression of a novel combination of fast and slow troponin T isoforms in rabbit extraocular muscles. J Muscle Res Cell Motil 9:241–247

Bronson DD, Schachat FH (1982) Heterogeneity of contractile proteins. Differences in tropomyosin in fast, mixed, and slow skeletal muscles of the rabbit. J Biol Chem 257:3937–3944

Brooke MH, Kaiser KK (1970a) Three "myosin adenosine triphosphatase" systems: the nature of their pH lability and sulfhydryl dependence. J Histochem Cytochem 18:670–672

Brooke MH, Kaiser KK (1970b) Muscle fiber types: how many and what kind? Arch Neurol 23:369–379

Brown HR, Schachat FH (1985) Renaturation of skeletal muscle tropomyosin: implications for in vivo assembly. Proc Natl Acad Sci USA 82:2359–2363

Brown JMC, Henriksson J, Salmons S (1989) Restoration of fast muscle characteristics following cessation of chronic stimulation: physiological, histochemical and metabolic changes during slow-to-fast transformation. Proc R Soc Lond [Biol] 235:321–346

Bruggmann S, Jenny E (1975) The immunological specificity of myosin from cross striated muscles as revealed by quantitative microcomplement fixation and enzyme inhibition by antisera. Biochim Biophys Acta 412:39–50

Buchegger A, Nemeth PM, Pette D, Reichmann H (1984) Effects of chronic stimulation on the metabolic heterogeneity of the fibre population in rabbit tibialis anterior muscle. J Physiol (Lond) 350:109–119

Buchthal F, Schmalbruch H (1980) Motor unit of mammalian muscle. Physiol Rev 60:90–142

Buckingham ME, Alonso S, Barton P, Bugaiski G, Cohen A, Daubas A, Minty A, Robert B, Weydert A (1984) Actin and myosin coding sequences in skeletal muscle development. Exp Biol Med 9:228–334

Buckingham M, Alonso S, Barton P, Cohen A, Daubas P, Garner I, Robert B, Weydert A (1986) Actin and myosin multigene families: their expression during the formation and maturation of striated muscle. Am J Med Genet 25:623–634

Burke RE (1981) Motor units: anatomy physiology and functional organization. In: Brookhart JM, Mountcastle VB (eds) Handbook of physiology, Sect 1, Vol II: The nervous system, Am Physiol Soc, Bethesda, pp 345–422

Burke RE, Levine DN, Zajac FE, Tsairis P, Engel WK (1971) Mammalian motor units: physiological-histochemical correlation in three types in cat gastrocnemius. Science 174:709–712

Burke RE, Levine DN, Tsairis P, Zajac FE (1973) Physiological types and histochemical profiles in motor units of the cat gastrocnemius. J Physiol (Lond) 234:723–748

Burke RE, Levine DN, Salcman M, Tsairis P (1974) Motor units in cat soleus muscle: physiological, histochemical and morphological characteristics. J Physiol (Lond) 238:503–514

Burleigh IG, Schimke RT (1969) The activities of some enzymes concerned with energy metabolism in mammalian muscles of differing pigmentation. Biochem J 113:157–166

Butler-Browne GS, Whalen RG (1984) Myosin isozyme transitions occuring during the postnatal development of the rat soleus muscle. Dev Biol 102:324–334

Butler-Browne GS, Eriksson P-O, Laurent C, Thornell LE (1988) Adult human masseter muscle fibers express myosin isozymes characteristic of development. Muscle Nerve 11:610–620

Callaway JE, Bechtel PJ (1981) C-protein from rabbit soleus (red) muscle. Biochem J 195:463–469

Carraro U, Catani C (1983) A sensitive SDS-PAGE method separating myosin heavy chain isoforms of rat skeletal muscles reveals the heterogeneous nature of the embryonic myosin. Biochem Biophys Res Commun 116:793–802

Carraro U, Dalla Libera L, Catani C (1981 a) Myosin light chains of avian and mammalian slow muscles: evidence of intraspecific polymorphism. J Muscle Res Cell Motil 2:335–342

Carraro U, Catani C, Dalla Libera L, Vascon M, Zanella G (1981 b) Differential distribution of tropomyosin subunits in fast and slow rat muscles and its changes in long-term denervated hemidiaphragm. FEBS Lett 128:233–236

Carraro U, Dalla Libera L, Catani C (1983) Myosin light and heavy chains in muscle regenerating in absence of the nerve: transient appearance of the embryonic light chain. Exp Neurol 79:106–117

Celio MR, Heizmann CW (1982) Calcium-binding protein parvalbumin is associated with fast contracting muscle fibres. Nature 297:504–506

Cerny LC, Bandman E (1987) Expression of myosin heavy chain isoforms in regenerating myotubes of innervated and denervated chicken pectoral muscle. Dev Biol 119:350–362

Chen WYJ, Dhoot GK, Perry SV (1986) Characterization and fibre type distribution of a new myofibrillar protein of molecular weight 32 kDa. J Muscle Res Cell Motil 7:517–526

Chi MM-Y, Hintz CS, Coyle EF, Wade HM III, Ivy JL, Nemeth PM, Holloszy JO (1983) Effects of detraining on enzymes of energy metabolism in individual human muscle fibers. Am J Physiol 244:C276–C287

Clarke FM, Lovell SJ, Masters CJ, Winzor DJ (1976) Beef muscle troponin: evidence for multiple forms of troponin-T. Biochim Biophys Acta 427:617–626

Close R (1967) Properties of motor units in fast and slow skeletal muscles of the rat. J Physiol (Lond) 193:45–55

Close RI (1972) Dynamic properties of mammalian skeletal muscles. Physiol Rev 52:129–197

Collins JH (1974) Homology of myosin light chains, troponin-C and parvalbumins deduced from comparison of their amino acid sequences. Biochem Biophys Res Commun 58:301–308

Collins JH, Theibert JL, Dalla Libera L (1986) Amino acid sequence of rabbit ventricular myosin light chain-2: identity with the slow skeletal muscle isoform. Biosci Rep 6:655–661

Crabtree B, Newsholme EA (1972) The activities of phosphorylase, hexokinase, phosphofructokinase, lactate dehydrogenase and the glycerol 3-phosphate dehydrogenases in muscles from vertebrates and invertebrates. Biochem J 126:49–58

Craig RW, Offer G (1976) The location of C-protein in rabbit skeletal muscle. Proc R Soc Lond [Biol] 192:451–461

Crow MT (1987) The determinants of muscle fiber type during embryonic development. Am Zool 27:1043–1053

Cummins P, Perry SV (1973) The subunits and biological activity of polymorphic forms of tropomyosin. Biochem J 133:765–777

Cummins P, Perry SV (1974) Chemical and immunochemical characteristics of tropomyosins from striated and smooth muscle. Biochem J 141:43–49

Cummins P, Price KM (1980) Identification of a foetal myosin light chain in the human ventricle. J Muscle Res Cell Motil 1:482–483

Dabrowska R, Dydynska M, Szpacenko A, Drabikowski W (1973) Comparative studies on the composition and properties of troponin from fast, slow and cardiac muscles. Int J Biochem 4:189–194

d'Albis A, Gratzer WB (1973) Electrophoretic examination of native myosin. FEBS Lett 29:292–296

d'Albis A, Pantaloni C, Béchet J-J (1979) An electrophoretic study of native myosin isozymes and of their subunit content. Eur J Biochem 99:261–272

d'Albis A, Pantolini C, Béchet J-J (1982) Myosin isoenzymes in adult rat skeletal muscles. Biochimie 64:399–404

d'Albis A, Janmot C, Béchet J-J (1986) Comparison of myosins from the masseter muscle of adult rat, mouse and guinea-pig. Persistence of neonatal-type isoforms in the murine muscle. Eur J Biochem 156:291–296

d'Albis A, Couteaux R, Janmot C, Roulet A, Mira J-C (1988) Regeneration after cardiotoxin injury of innervated and denervated slow and fast muscles of mammals. Myosin isoform analysis. Eur J Biochem 174:103–110

d'Albis A, Couteaux R, Janmot C, Roulet A (1989a) Specific programs of myosin expression in the postnatal development of rat muscles. Eur J Biochem 183:583–590

d'Albis A, Couteaux R, Janmot C, Mira JC (1989b) Myosin isoform transitions in regeneration of fast and slow muscles during postnatal development of the rat. Dev Biol 135:320–325

Dalla Libera L (1981) Myosin heavy chains in fast skeletal muscle of chick embryo. Experientia 37:1268–1270

Dalla Libera L (1988) A comparative study of chicken ventricular and slow skeletal myosin light chains. Cell Biol Int Rep 12:1089–1098

Dalla Libera L, Sartori S, Schiaffino S (1979) Comparative analysis of chicken atrial and ventricular myosin. Biochim Biophys Acta 581:283–294

Dalla Libera L, Sartore S, Pierobon-Bormioli S, Schiaffino S (1980) Fast-white and fast-red isomyosins in guinea pig muscles. Biochem Biophys Res Commun 96:1662–1670

Dalrymple RH, Cassens RG, Kastenschmidt LL (1974) Glycolytic enzyme activity in developing red and white muscle. J Cell Physiol 83:251–257

Damiani E, Betto R, Salvatori S, Volpe P, Salviati G, Margreth A (1981) Polymorphism of sarcoplasmic reticulum adenosine triphosphatase of rabbit skeletal muscle. Biochem J 197:245–248

Damiani E, Salvatori S, Zorzato F, Margreth A (1986) Characteristics of skeletal muscle calsequestrin: comparison with mammalian, amphibian and avian muscles. J Muscle Res Cell Motil 7:435–445

Dangain J, Pette D, Vrbová G (1987) Developmental changes in succinate dehydrogenase activity in muscle fibers from normal and dystrophic mice. Exp Neurol 95:224–234

Danieli-Betto D, Zerbato E, Betto R (1986) Type 1, 2A, and 2B myosin heavy chain electrophoretic analysis of rat muscle fibers. Biochem Biophys Res Commun 138:981–987

Davey DF, Wong SYP (1980) Morphometric analysis of rat extensor digitorum longus and soleus muscles. Aust J Exp Biol Med Sci 58:213–230

Davison MD, Critchley DR (1988) α-Actinins and the DMD protein contain spectrin-like repeats. Cell 52:159–160

Dawson DM, Romanul FCA (1964) Enzymes in Muscle. II. Histochemical and quantitative studies. Arch Neurol 11:369–378

Dechesne CA, Bouvagnet P, Walzthoeny D, Léger JJ (1987) Visualization of cardiac ventricular myosin heavy chain homodimers and heterodimers by monoclonal antibody epitope mapping. J Cell Biol 105:3031–3037

DeFoor PH, Levitsky D, Biryukova T, Fleischer S (1980) Immunological dissimilarity of the calcium pump protein of skeletal and cardiac muscle sarcoplasmic reticulum. Arch Biochem Biophys 200:196–205

De Meis L (1981) The sarcoplasmic reticulum. Transport and energy transduction. In: Bittar EE (ed) The sarcoplasmic reticulum: transport and energy transduction. Wiley, New York

Dennis JE, Shimizu T, Reinach FC, Fischman DA (1984) Localization of C-protein isoforms in chicken skeletal muscle: ultrastructural detection using monoclonal antibodies. J Cell Biol 98:1514–1522

Denny-Brown DE (1929) The histological features of striped muscle in relation to its functional activity. Proc R Soc Lond [Biol] 104:371–410

Dhoot GK (1988) Identification and distribution of the fast class of troponin T in the adult and developing avian skeletal muscle. J Muscle Res Cell Motil 9:446–455

Dhoot GK, Perry SV (1979) Distribution of polymorphic forms of troponin components and tropomyosin in skeletal muscle. Nature 278:714–718

Dhoot GK, Perry SV (1980) Factors determining the expression of the genes controlling the synthesis of the regulatory proteins in striated muscle. In: Pette D (ed) Plasticity of muscle. de Gruyter, Berlin, pp 255–267

Dhoot GK, Gell PGH, Perry SV (1978) Localization of different forms of troponin I in skeletal and cardiac muscle cells. Exp Cell Res 117:357–370

Dhoot GK, Frearson N, Perry SV (1979) Polymorphic forms of troponin T and troponin C and their localization in striated muscle cell types. Exp Cell Res 122:339–350

Dhoot GK, Hales MC, Grail BM, Perry SV (1985) The isoforms of C protein and their distribution in mammalian skeletal muscle. J Muscle Res Cell Motil 6:487–505

Dölken G, Pette D (1974) Turnover of several glycolytic enzymes in rabbit heart, soleus muscle and liver. Hoppe-Seyler's Z Physiol Chem 355:289–299

Doriguzzi C, Mongini T, Palmucci L, Schiffer D (1983) A new method for myofibrillar Ca^{++}-ATPase reaction based on the use of metachromatic dyes: its advantages in muscle fibre typing. Histochemistry 79:289–294

Dow J, Stracher A (1971) Identification of the essential light chains of myosin. Proc Natl Acad Sci USA 68:1107–1110

Dubowitz V, Brooke MH (1973) Muscle biopsy: a modern approach. In: Walton JN (ed) Major problems in neurology. Saunders, London

Dubowitz V, Pearse AGE (1960) A comparative histochemical study of oxidative enzyme and phosphorylase activity in skeletal muscle. Histochemie 2:105–117

Dulhunty AF, Banyard MRC, Medveczky CJ (1987) Distribution of calcium ATPase in the sarcoplasmic reticulum of fast- and slow-twitch muscles determined with monoclonal antibodies. J Membr Biol 99:79–92

Dux L, Martonosi A (1984) Membrane crystals of Ca^{2+}-ATPase in sarcoplasmic reticulum of fast and slow skeletal and cardiac muscles. Eur J Biochem 141:43–49

Eaton L, Pepe FA (1972) Two components isolated from chicken breast muscle. J Cell Biol 55:681–695

Ebashi S (1972) Separation of troponin into its three components. J Biochem (Tokyo) 72:787–790

Ebashi S, Ebashi F (1965) α-actinin, a new structural protein from skeletal muscle. I. Preparation and action on actomyosin-ATP interaction. J Biochem (Tokyo) 58:7–12

Eddinger TJ, Moss RL (1987) Mechanical properties of skinned single fibers of identified types from rat diaphragm. Am J Physiol 252:C210–C218

Eddinger TJ, Moss RL, Cassens RG (1985) Myosin-ATPase fibre typing of chemically skinned muscle fibres. Histochem J 17:1021–1026

Edgerton VR, Simpson DR (1969) The intermediate muscle fiber of rats and guinea pigs. J Histochem Cytochem 17:828–838

Edgerton VR, Simpson DR (1971) Dynamic and metabolic relationships in the rat extensor digitorum longus muscle. Exp Neurol 30:374–376

Edgerton VR, Roy RR, Bodine-Fowler SC, Pierotti DJ, Unguez GA, Martin TP, Jiang B, Chalmers RS (1990) Motoneurons − muscle fiber connectivity and interdependence. In: Pette D (ed) The dynamic state of muscle fibers. de Gruyter, Berlin, pp 217−231

Edström L, Kugelberg E (1968) Histochemical composition, distribution of fibres and fatiguability of single motor units. Anterior tibial muscle of the rat. J Neurol Neurosurg Psychiatry 31:424−433

Eisenberg BR (1974) Quantitative ultrastructural analysis of adult mammalian skeletal muscle fibers. In: Milhorat AT (ed) Exploratory concepts in muscular dystrophy, vol. 2. Excerpta Medica, Amsterdam, pp 258−270

Eisenberg BR (1983) Quantitative ultrastructure of mammalian skeletal muscle. In: Peachey LD, Adrian RH, Geiger SR (eds) Handbook of physiology, Sect 10: Skeletal muscle, Williams and Wilkins, Baltimore Md, pp 73−112

Eisenberg BR, Kuda AM (1975) Stereological analysis of mammalian skeletal muscle. II. White vastus muscle of the adult guinea pig. J Ultrastruct Res 51:176−187

Eisenberg BR, Kuda AM (1976) Discrimination between fiber populations in mammalian skeletal muscle by using ultrastructural parameters. J Ultrastruct Res 54:76−88

Eisenberg BR, Salmons S (1981) The reorganization of subcellular structure in muscle undergoing fast-to-slow type transformation. A stereological study. Cell Tissue Res 220:449−471

Eisenberg BR, Kuda AM, Peter JB (1974) Stereological analysis of mammalian skeletal muscle. I. Soleus muscle of the adult guinea pig. J Cell Biol 60:732−754

Eisenberg BR, Brown JMC, Salmons S (1984) Restoration of fast muscle characteristics following cessation of chronic stimulation. Cell Tissue Res 238:221−230

Eisenberg BR, Dix DG, Lin ZW, Wenderoth MP (1987) Relationship of membrane systems in muscle to isomyosin content. Can J Physiol Pharmacol 65:598−605

El-Saleh SC, Warber KD, Potter JD (1986) The role of tropomyosin-troponin in the regulation of skeletal muscle contraction. J Muscle Res Cell Motil 7:387−404

Elzinga M, Collins JH, Kuehl WM, Adelstein RS (1973) Complete amino-acids sequence of actin of rabbit skeletal muscle. Proc Natl Acad Sci USA 70:2687−2691

Emerson CP Jr. (1987) Molecular genetics of myosin. Ann Rev Biochem 56:695−726

Engel WK (1962) The essentiality of histo- and cytochemical studies of skeletal muscle in the investigation of neuromuscular disease. Neurology 12:778−784

Eppenberger HM, Perriard J-C, Rosenberg UB, Strehler EE (1981) The M_r 165,000 M-protein myomesin: a specific protein of cross-striated muscle cells. J Cell Biol 89:185−193

Essén B, Jansson E, Henriksson J, Taylor AW, Saltin B (1975) Metabolic characteristics of fibre types in human skeletal muscle. Acta Physiol Scand 95:153−165

Essén-Gustavsson B (1986) Activity- and inactivity-related muscle adaptation in the animal kingdom. In: Saltin B (ed) Biochemistry of exercise, vol 6. Human Kinetics Publishers, Champaign, pp 435−444

Essén-Gustavsson B, Henriksson J (1984) Enzyme levels in pools of microdissected human muscle fibres of identified type. Adaptive response to exercise. Acta Physiol Scand 120:505−515

Eusebi F, Miledi R, Takahashi T (1980) Calcium transients in mammalian muscles. Nature 284:560−561

Feramisco JR, Burridge K (1980) A rapid purification of α-actinin, filamin and a 130000-dalton protein from smooth muscle. J Biol Chem 255:1194−1199

Ferguson DG, Franzini-Armstrong C (1988) The Ca^{2+} ATPase content of slow and fast twitch fibers of guinea pig. Muscle Nerve 11:561−570

Fiehn W, Peter JB (1971) Properties of the fragmented sarcoplasmic reticulum from fast twitch and slow twitch muscle. J Clin Invest 50:570−573

Fitzsimons RB, Hoh JFY (1981) Isomyosin in human type 1 and type 2 skeletal muscle fibres. Biochem J 193:229−233

Fitzsimons RB, Hoh JFY (1983) Myosin isoenzymes in fast-twitch and slow-twitch muscles of normal and dystrophic mice. J Physiol (Lond) 343:539−550

Flicker PF, Wallimann T, Vibert P (1983) Electron microscopy of scallop myosin. Location of regulatory light chains. J Mol Biol 169:723−741

Fliegel L, Ohnishi M, Carpenter MR, Khanna VK, Reithmeier RA, MacLennan DH (1987) Amino acid sequence of rabbit fast-twitch skeletal muscle calsequestrin deduced from cDNA and peptide sequencing. Proc Natl Acad Sci USA 84:1167−1171

Fliegel L, Leberer E, Green MN, MacLennan DH (1989) The fast-twitch muscle calse-questrin isoform predominates in rabbit slow-twitch soleus muscle. FEBS Lett 242:297−300

Franzini-Armstrong C (1984) Freeze-fracture of frog slow tonic fibers. Structure of surface and internal membranes. Tissue Cell 16:647−664

Franzini-Armstrong C, Ferguson DG, Champ C (1988) Discrimination between fast- and slow-twitch fibres of guinea pig skeletal muscle using the relative surface density of junctional transverse tubule membrane. J Muscle Res Cell Motil 9:403−414

Frémont P, Lazure C, Tremblay RR, Chrétien M, Rogers PA (1987) Regulation of carbonic anhydrase III by thyroid hormone: opposite modulation in slow- and fast-twitch skeletal muscle. Biochem Cell Biol 65:790−797

Frémont P, Charest PM, Coté C, Rogers PA (1988) Carbonic anhydrase III in skeletal muscle fibers: an immunocytochemical and biochemical study. J Histochem Cytochem 36:775−782

Fujii J, Lytton J, Tada M, MacLennan DH (1988) Rabbit cardiac and slow-twitch muscle express the same phospholamban gene. FEBS Lett 227:51−55

Funatsu T, Asami Y, Ishiwata S (1988) β-Actinin: a capping protein at the pointed end of thin filaments in skeletal muscle. J Biochem (Tokyo) 103:61−71

Gagnon J, Ho-Kim MA, Champagne C, Tremblay RR, Rogers PA (1985) Modulation of a major 30-kDa skeletal muscle protein by thyroid hormone. FEBS Lett 180:335−340

Gahlmann R, Wade R, Gunning P, Kedes L (1988) Differential expression of slow and fast skeletal muscle troponin C. Slow skeletal muscle troponin C is expressed in human fibroblasts. J Mol Biol 201:379−391

Gambke B, Rubinstein NA (1984) A monoclonal antibody to the embryonic myosin heavy chain of rat skeletal muscle. J Biol Chem 259:12092−12100

Gambke B, Lyons GE, Haselgrove J, Kelly AM, Rubinstein NA (1983) Thyroidal and neural control of myosin transitions during development of rat fast and slow muscles. FEBS Lett 156:335−339

Garrels JA, Gibson W (1976) Identification and characterization of multiple forms of actin. Cell 9:793−805

Gauthier GF (1969) On the relationship of ultrastructural and cytochemical features to color in mammalian skeletal muscle. Z Zellforsch 95:462−482

Gauthier GF (1974) Some ultrastructural and cytochemical features of fiber populations in the soleus muscle. Anat Rec 180:551−564

Gauthier GF (1986) Skeletal muscle fiber types. In: Engel AG, Banker BG (eds) Myology, vol 1. McGraw-Hill, New York, part 1, chap 8, pp 255−283

Gauthier GF (1987) Vertebrate muscle fiber types and neuronal regulation of myosin expression. Am Zool 27:1033−1042

Gauthier GF (1990) Differential distribution of myosin isoforms among the myofibrils of individual developing muscle fibers. J Cell Biol 110:693−701

Gauthier GF, Hobbs AW (1986) Freeze-fractured sarcoplasmic reticulum in adult and embryonic fast and slow muscles. J Muscle Res Cell Motil 7:122−133

Gauthier GF, Lowey S (1977) Polymorphism of myosin among skeletal muscle fiber types. J Cell Biol 74:760−779

Gauthier GF, Lowey S (1979) Distribution of myosin isoenzymes among skeletal muscle fiber types. J Cell Biol 81:10−25

Gauthier GF, Lowey S, Hobbs AW (1978) Fast and slow myosin in developing muscle fibers. Nature 274:25–29

Gauthier GF, Burke RE, Lowey S, Hobbs AW (1983) Myosin isozymes in normal and cross-reinnervated cat skeletal muscle fibers. J Cell Biol 97:756–771

Gazith J, Himmelfarb S, Harrington WF (1970) Studies on the subunit structure of myosin. J Biol Chem 245:15–22

Gershman LC, Stracher A, Dreizen P (1969) Subunit structure of myosin. III. A proposed model for rabbit skeletal myosin. J Biol Chem 244:2726–2736

Gillis JM (1985) Relaxation of vertebrate skeletal muscle. A synthesis of the biochemical and physiological approaches. Biochim Biophys Acta 811:97–145

Gillis JM, Thomason D, Lefèvre J, Kretsinger RH (1982) Parvalbumins and muscle relaxation: a computer simulation study. J Muscle Res Cell Motil 3:377–398

Gollnick PD, Matoba H (1984) Identification of fiber types in rat skeletal muscle based on the sensitivity of myofibrillar actomyosin ATPase to copper. Histochemistry 81:379–383

Gollnick PD, Armstrong RB, Saubert IV CW, Piehl K, Saltin B (1972) Enzyme activity and fiber composition in skeletal muscle of untrained and trained men. J Appl Physiol 33:312–319

Gollnick PD, Armstrong RB, Saltin B, Saubert IV CW, Sembrowich WL, Shepherd RE (1973) Effect of training on enzyme activity and fiber composition of human skeletal muscle. J Appl Physiol 34:107–111

Gollnick PD, Parsons D, Oakley CR (1983) Differentiation of fiber types in skeletal muscle from the sequential inactivation of myofibrillar actomyosin ATPase during acid preincubation. Histochemistry 77:543–555

Gorza L (1990) Identification of a novel type 2 fiber population in mammalian skeletal muscle by combined use of histochemical myosin ATPase and anti-myosin monoclonal antibodies. J Histochem Cytochem 38:257–265

Greaser ML, Gergely J (1973) Purification and properties of the components from troponin. J Biol Chem 248:2125–2133

Greaser ML, Moss RL, Reiser PJ (1988) Variations in contractile properties of rabbit single muscle fibres in relation to troponin T isoforms and myosin light chains. J Physiol (Lond) 406:85–98

Green HJ, Thomson JA, Daub WD, Houston ME, Ranney DA (1979) Fiber composition, fiber size and enzyme activities in vastus lateralis of elite athletes involved in high intensity exercise. Eur J Appl Physiol 41:109–117

Green HJ, Reichmann H, Pette D (1982) A comparison of two ATPase based schemes for histochemical muscle fibre typing in various mammals. Histochemistry 76:21–31

Green HJ, Reichmann H, Pette D (1983) Fibre type specific transformations in the enzyme activity pattern of rat vastus lateralis muscle by prolonged endurance training. Pflügers Arch 399:216–222

Green HJ, Klug GA, Reichmann H, Seedorf U, Wiehrer W, Pette D (1984) Exercise-induced fibre type transitions with regard to myosin, parvalbumin, and sarcoplasmic reticulum in muscles of the rat. Pflügers Arch 400:432–438

Gregory P, Low RB, Stirewalt WS (1986) Changes in skeletal-muscle myosin isoenzyme with hypertrophy and exercise. Biochem J 238:55–63

Gregory P, Low RB, Stirewalt WS (1987) Fractional synthesis rates in vivo of skeletal-muscle myosin isoenzymes. Biochem J 245:133–137

Gröschel-Stewart U, Doniach D (1969) Immunological evidence for human myosin isoenzymes. Immunology 17:991–994

Gröschel-Stewart U, Drenckhahn D (1982) Muscular and cytoplasmic contractile proteins. Coll Relat Res 2:381–463

Grove BK, Cerny L, Perriard J-C, Eppenberger HM (1985) Myomesin and M-protein: expression of two M-band proteins in pectoral muscle and heart during development. J Cell Biol 101:1413–1421

Grove BK, Cerny L, Perriard J-C, Eppenberger HM, Thornell L-E (1989) Fiber type-specific distribution of M-band proteins in chicken muscle. J Histochem Cytochem 37:447–454

Grützner P (1883) Zur Physiologie und Histologie der Skelettmuskeln. Breslauer Ärztl Z 5:257–258

Gundersen K, Leberer E, Lömo T, Pette D, Staron RS (1988) Fibre types, calcium-sequestering proteins and metabolic enzymes in denervated and chronically stimulated muscles of the rat. J Physiol (Lond) 398:177–189

Gunning P, Ponte P, Blau H, Kedes L (1983) α-skeletal and α-cardiac actin genes are co-expressed in adult human skeletal muscle and heart. Mol Cell Biol 3:1985–1995

Guth L, Samaha FJ (1969) Qualitative differences between actomyosin ATPase of slow and fast mammalian muscle. Exp Neurol 25:138–152

Guth L, Samaha FJ (1972) Erroneous interpretations which may result from application of the "myofibrillar ATPase" histochemical procedure to developing muscle. Exp Neurol 34:465–475

Guth L, Yellin H (1971) The dynamic nature of the so-called "fiber types" of mammalian skeletal muscle. Exp Neurol 31:277–300

Häggmark T, Jansson E, Eriksson E (1981) Fiber type area and metabolic potential of the thigh muscle in man after knee surgery and immobilization. Int J Sports Med 2:12–17

Hamm TM, Nemeth PM, Solanki L, Gordon DA, Reinking RM, Stuart DG (1988) Association between biochemical and physiological properties in single motor units. Muscle Nerve 11:245–254

Hammonds RG Jr (1987) Protein sequence of DMD gene is related to actin-binding domain of α-actinin. Cell 51:1

Harigaya S, Ogawa Y, Sugita H (1968) Calcium binding activity of microsomal fraction of rabbit red muscle. J Biochem (Tokyo) 63:324–331

Harris DA, Falls DL, Fischbach GD (1989) Differential activation of myotube nuclei following exposure to an acetylcholine receptor-inducing factor. Nature 337:173–176

Härtner K-T, Pette D (1988) Low-frequency stimulation-induced fast to slow transitions of troponin subunit isoforms in rabbit skeletal muscle. In: Carraro U (ed) Sarcomeric and non-sarcomeric muscles: basic and applied research for the 90s, CLEUP, Padova, pp 367–371

Härtner K-T, Pette D (1990) Fast and slow isoforms of troponin I and troponin C. Normal rabbit muscles and effects of chronic stimulation. Eur J Biochem 188:261–267

Härtner K-T, Kirschbaum BJ, Pette D (1989) The multiplicity of troponin T isoforms. Normal rabbit muscles and effects of chronic stimulation. Eur J Biochem 179:31–38

Hasselbach W (1964) Relaxing factor and the relaxation of muscle. Prog Biophys Mol Biol 14:167–222

Hayward LJ, Schwartz RJ (1986) Sequential expression of chicken actin genes during myogenesis. J Cell Biol 102:1485–1493

Heeley DH, Dhoot GK, Frearson N, Perry SV, Vrbová G (1983) The effect of cross-innervation on the tropomyosin composition of rabbit skeletal muscle. FEBS Lett 152:282–286

Heeley DH, Dhoot GK, Perry SV (1985) Factors determining the subunit composition of tropomyosin in mammalian skeletal muscle. Biochem J 226:461–468

Heilmann C, Pette D (1979) Molecular transformations in sarcoplasmic reticulum of fast-twitch muscle by electro-stimulation. Eur J Biochem 93:437–446

Heilmann C, Brdiczka D, Nickel E, Pette D (1977) ATPase activities, Ca^{2+} transport and phosphoprotein formation in sarcoplasmic reticulum subfractions of fast and slow rabbit muscles. Eur J Biochem 81:211–222

Heilmann C, Müller W, Pette D (1981) Correlation between ultrastructural and functional changes in sarcoplasmic reticulum during chronic stimulation of fast muscle. J Membrane Biol 59:143–149

Heizmann CW (1984) Parvalbumin, an intracellular calcium-binding protein; distribution, properties and possible roles in mammalian cells. Experientia 40:910–921

Heizmann CW, Berchtold MW, Rowlerson AM (1982) Correlation of parvalbumin concentration with relaxation speed in mammalian muscles. Proc Natl Acad Sci USA 79:7243–7247

Heizmann CW, Celio MR, Billeter R (1983) A new myofibrillar protein characteristic of type I human skeletal muscle fibres. Eur J Biochem 132:657–662

Henneman E, Olson CB (1965) Relations between structure and function in the design of skeletal muscles. J Neurophysiol 28:581–598

Henriksson J, Chi MM-Y, Hintz CS, Young DA, Kaiser KK, Salmons S, Lowry OH (1986) Chronic stimulation of mammalian muscle: changes in enzymes of six metabolic pathways. Am J Physiol 251:C614–C632

Hikida RS (1978) Z-line extraction: comparative effects in avian skeletal muscle fiber types. J Ultrastruct Res 65:266–278

Hill C, Weber K (1986) Monoclonal antibodies distinguish titins from heart and skeletal muscle. J Cell Biol 102:1099–1108

Hintz CS, Lowry CV, Kaiser KK, McKee D, Lowry OH (1980) Enzyme levels in individual rat muscle fibers. Am J Physiol 239:C58–C65

Hintz CS, Chi MM-Y, Fell RD, Ivy JL, Kaiser KK, Lowry CV, Lowry OH (1982) Metabolite changes in individual rat muscle fibers during stimulation. Am J Physiol 242:C218–C228

Hintz CS, Coyle EF, Kaiser KK, Chi MM-Y, Lowry OH (1984a) Comparison of muscle fiber typing by quantitative enzyme assays and by myosin ATPase staining. J Histochem Cytochem 32:655–660

Hintz CS, Chi MM-Y, Lowry OH (1984b) Heterogeneity in regard to enzymes and metabolites within individual muscle fibers. Am J Physiol 246:C288–C292

Hirsch HE, Parks ME, Blanco CE, Simpson DR (1982) The ratio of 3-hydroxyacyl-CoA dehydrogenase to lipoamide dehydrogenase activity in individual muscle fibers: mitochondrial specialization for source of energy. J Neurosci Res 8:7–12

Hoffman EP, Knudsdon CM, Campbell KP, Kunkel LM (1987) Subcellular fractionation of dystrophin to the triads of skeletal muscle. Nature 330:754–758

Hoffman EP, Watkins SC, Slayter HS, Kunkel LM (1989) Detection of a specific isoform of alpha-actinin with antisera directed against dystrophin. J Cell Biol 108:503–510

Hofmann S, Düsterhöft S, Pette D (1988) Six myosin heavy chain isoforms are expressed during chick breast muscle development. FEBS Lett 238:245–248

Hoh JFY (1975) Neural regulation of mammalian fast and slow muscle myosins: an electrophoretic analysis. Biochemistry 14:742–747

Hoh JFY (1978) Light chain distribution of chicken skeletal muscle myosin isoenzymes. FEBS Lett 90:297–300

Hoh JFY, Yeoh GPS (1979) Rabbit skeletal myosin isoenzymes from fetal, fast-twitch and slow-twitch muscles. Nature 280:321–322

Hoh JFY, McGrath PA, White RI (1976) Electrophoretic analysis of multiple forms of myosin in fast-twitch and slow-twitch muscles of the chick. Biochem J 157:87–95

Hoh JFY, Yeoh GPS, Thomas MAW, Higginbottom L (1979) Structural differences in the heavy chains of rat ventricular myosin isozymes. FEBS Lett 97:330–334

Hoh JFY, Hughes S, Chow C, Hale PT, Fitzsimons RB (1988) Immunocytochemical and electrophoretic analyses of changes in myosin gene expression in cat posterior temporalis muscle during postnatal development. J Muscle Res Cell Motil 9:48–58

Holloszy JO, Coyle EF (1984) Adaptations of skeletal muscle to endurance exercise and their metabolic consequences. J Appl Physiol 56:831–838

Hood DA, Pette D (1989) Chronic long-term electrostimulation creates a unique metabolic enzyme profile in rabbit fast-twitch muscle. FEBS Lett 247:471–474

Hood DA, Zak R, Pette D (1989) Chronic stimulation of rat skeletal muscle induces coordinate increases in mitochondrial and nuclear mRNAs of cytochrome c oxidase subunits. Eur J Biochem 179:275–280

Hoppeler H (1986) Exercise-induced ultrastructural changes in skeletal muscle. Int J Sports Med 7:187–204

Hoppeler H (1990) The range of mitochondrial adaptation in muscle fibers. In: Pette D (ed) The dynamic state of muscle fibers. de Gruyter, Berlin, pp 567–586

Hoppeler H, Claassen H, Howald H, Straub R (1983) Correlated histochemistry and morphometry in equine skeletal muscle. In: Snow DH, Persson SGB, Rose RJ (eds) Equine exersice physiology, Granta, Cambridge, pp 184–192

Hoppeler H, Hudlická O, Uhlmann E (1987) Relationship between mitochondria and oxygen consumption in isolated cat muscles. J Physiol (Lond) 385:661–675

Horowits R, Podolsky RJ (1987) The positional stability of thick filaments in activated skeletal muscle depends on sarcomere length: evidence for the role of titin filaments. J Cell Biol 105:2217–2223

Horowits R, Kempner ES, Bisher ME, Podolsky RJ (1986) A physiological role for titin and nebulin in skeletal muscle. Nature 323:160–164

Houston ME, Green HJ, Stull JT (1985) Myosin light chain phosphorylation and isometric twitch potentiation in intact human muscle. Pflügers Arch 403:348–352

Houston ME, Lingley MD, Stuart DS, Grange RW (1987) Myosin light chain phosphorylation in intact human muscle. FEBS Lett 219:469–471

Howald H (1982) Training-induced morphological and functional changes in skeletal muscle. Int J Sports Med 3:1–12

Howald H, Hoppeler H, Claasen H, Mathieu O, Straub R (1985) Influences of endurance training on the ultrastructural composition of the different muscle fiber types in humans. Pflügers Arch 403:369–376

Hudlická O, Pette D, Staudte H (1973) The relation between blood flow and enzymatic activities in slow and fast muscles during development. Pflügers Arch 343:341–356

Hughes VM (1986) A new histochemical method for magnesium actomyosin adenosine triphosphatase at physiological pH. Stain Technol 61:201–214

Hughes SM, Blau HM (1990) Regulation or regional specialization in muscle fibres. In: Pette D (ed) The dynamic state of muscle fibers. de Gruyter, Berlin, pp 265–277

Illg D, Pette D (1979) Turnover rates of hexokinase I, phosphofructokinase, pyruvate kinase and creatine kinase a slow-twitch soleus muscle and heart of the rabbit. Eur J Biochem 97:267–273

Ingjer F (1979) Effects of endurance training on muscle fibre ATPase activity, capillary supply and mitochondrial content in man. J Physiol (Lond) 294:419–432

Ishiura S, Takagi A, Nonaka I, Sugita H (1981) Heterogeneous expression of myosin light chain 1 in a human slow-twitch muscle fiber. J Biochem 90:279–282

Jansson E, Kaijser L (1977) Muscle adaptation to extreme endurance training in man. Acta Physiol Scand 100:315–324

Jansson E, Sjödin B, Tesch P (1978) Changes in muscle fibre type distribution in man after physical training – a sign of fibre type transformation? Acta Physiol Scand 104:235–237

Jansson E, Sylven C, Nordevang E (1982) Myoglobin in the quadriceps femoris muscle of competitive cyclists and untrained men. Acta Physiol Scand 114:627–629

Jean DH, Guth L, Albers RW (1973) Neural regulation of the structure of myosin. Exp Neurol 38:458–471

Jeffery S, Carter ND, Smith A (1986) Immunocytochemical localization of carbonic anhydrase isozymes I, II, and III in rat skeletal muscle. J Histochem Cytochem 34:513–516

Jeffery S, Carter ND, Smith A (1987) Thyroidectomy significantly alters carbonic anhydrase III concentration and fiber distribution in rat muscle. J Histochem Cytochem 35:663–668

Jeffery S, Kelly CD, Carter N, Kaufmann M, Termin A, Pette D (1990) Chronic stimulation-induced effects point to a coordinated expression of carbonic anhydrase III and slow myosin heavy chain in skeletal muscle. FEBS Lett 262:225–227

Jolesz F, Sréter FA (1981) Development, innervation, and activity-pattern induced changes in skeletal muscle. Ann Rev Physiol 43:531–552

Jorgensen AO, Jones LR (1986) Localization of phospholamban in slow but not fast canine skeletal muscle fibers. J Biol Chem 261:3775–3781

Jorgensen AO, McGuffee LJ (1987) Immunoelectron microscopic localization of sarcoplasmic reticulum proteins in cryofixed, freeze-dried, and low temperature-embedded tissue. J Histochem Cytochem 35:723–732

Jorgensen AO, Kalnins V, MacLennan DH (1979) Localization of sarcoplasmic reticulum proteins in rat skeletal muscle by immunofluorescence. J Cell Biol 80:372–384

Jorgensen AO, Shen ACY, MacLennan DH, Tokuyasu KT (1982) Ultrastructural localization of the $Ca^{2+}+Mg^{2+}$-dependent ATPase of sarcoplasmic reticulum in rat skeletal muscle by immunoferritin labeling of ultrathin frozen sections. J Cell Biol 92:409–416

Jorgensen AO, Shen ACY, Campbell KP, MacLennan DH (1983) Ultrastructural localization of calsequestrin in rat skeletal muscle by immunoferritin labeling of ultrathin frozen sections. J Cell Biol 97:1573–1581

Jorgensen AO, Shen ACY, Campbell KP (1985) Ultrastructural localization of calsequestrin in adult rat atrial and ventricular muscle cells. J Cell Biol 101:257–268

Kaprielian Z, Fambrough DM (1987) Expression of fast and slow isoforms of the Ca^{2+}-ATPase in developing chick skeletal muscle. Dev Biol 124:490–503

Kardami E, Montarras D, Fiszman M (1983) Fast and slow chicken skeletal muscles contain different alpha and beta tropomyosins. Biochem Biophys Res Commun 110:147–154

Karpati G, Eisen AA, Carpenter S (1975) Subtypes of the histochemical type I muscle fibers. J Histochem Cytochem 23:89–91

Katoh T, Lowey S (1989) Mapping myosin light chains by immonolectron microscopy. Use of anti-fluorescyl antibodies as structural probes. J Cell Biol 109:1549–1560

Kaufmann M, Simoneau J-A, Veerkamp JH, Pette D (1989) Electrostimulation-induced increases in fatty acid-binding protein and myoglobin in rat fast-twitch muscle and comparison with tissue levels in heart. FEBS Lett 245:181–184

Kelly AM, Rubinstein NA (1980) Why are fetal muscles slow? Nature 288:266–269

Khan MA (1978) Histoenzymatic characterization of subtypes of types I fibres in fast muscles of rats. Histochemistry 55:129–138

Khan MA, Papadimitriou JM, Holt PG, Kakulas BA (1972) A calcium-citro-phosphate technique for the histochemical localization of myosin ATPase. Stain Technol 47:277–281

Khan MA, Papadimitriou JM, Kakulas BA (1974) The effect of temperature on the pH stability of myosin ATPase as demonstrated histochemically. Histochemistry 38:181–194

Kim DH, Witzman FA, Fitts RH (1981) A comparison of sarcoplasmic reticulum in fast and slow skeletal muscles using crude homogenate and isolated vesicles. Life Sci 28:2223–2229

Kirchberger MA, Tada M (1976) Effects of adenosine 3':5'-monophosphate-dependent protein kinase on sarcoplasmic reticulum isolated from cardiac and slow and fast contracting skeletal muscles. J Biol Chem 251:725–729

Kirschbaum BJ, Pette D (1988) Low-frequency stimulation of rat fast-twitch muscle induces rapid, reversible changes in myosin heavy chain expression. In: Carraro U (ed) Sarcomeric and non-sarcomeric muscles: basic and applied research prospects for the 90s. CLEUP, Padova, pp 337–342

Kirschbaum BJ, Simoneau J-A, Bär A, Barton PJR, Buckingham ME, Pette D (1989a) Chronic stimulation-induced changes of myosin light chains at the mRNA and protein levels in rat fast-twitch muscle. Eur J Biochem 179:23–29

Kirschbaum BJ, Simoneau J-A, Pette D (1989b) Dynamics of myosin expression during the induced transformation of adult rat fast-twitch muscle. In: Stockdale F, Kedes L (eds) Cellular and molecular biology of muscle development. Liss, New York, pp 461–469

Kirschbaum BJ, Schneider S, Izumo S, Mahdavi V, Nadal-Ginard B, Pette D (1990a) Rapid and reversible changes in myosin heavy chain expression in response to increased neuromuscular activity of rat fast-witch muscle. FEBS Lett 268:75–78

Kirschbaum BJ, Kucher H-B, Termin A, Kelly AM, Pette D (1990b) Antagonistic effects of chronic low-frequency stimulation and thyroid hormone on myosin expression in rat fast-twitch muscle. J Biol Chem 265:13974–13980

Klug G, Reichmann H, Pette D (1983a) Rapid reduction in parvalbumin concentration during chronic stimulation of rabbit fast twitch muscle. FEBS Lett 152:180–182

Klug G, Wiehrer W, Reichmann H, Leberer E, Pette D (1983b) Relationships between early alterations in parvalbumin, sarcoplasmic reticulum and metabolic enzymes in chronically stimulated fast twitch muscle. Pflügers Arch 399:280–284

Klug GA, Leberer E, Leisner E, Simoneau J-A, Pette D (1988) Relationship between parvalbumin content and the speed of relaxation in chronically stimulated rabbit fast-twitch muscle. Pflügers Arch 411:126–131

Knoll P (1891) Über protoplasmaarme und protoplasmareiche Musculatur. Denkschr Kais Akad Wiss Wien Math Naturwiss Cl 58:633–700

Kobayashi R, Itoh H, Tashima Y (1983) Polymorphism of α-actinin. Electorphoretic and immunological studies of rabbit skeletal muscle alpha-actinins. Eur J Biochem 133:607–611

Kobayashi R, Itoh H, Tashima Y (1984) Different muscle-specific forms of rabbit skeletal muscle α-actinin. Eur J Biochem 143:125–131

Koenig M, Monaco AP, Kunkel LM (1988) The complete sequence of dystrophin predicts a rod-shaped cytoskeletal protein. Cell 53:219–228

Korczak B, Zarain-Herzberg A, Brandl CJ, Ingles CJ, Green NM, MacLennan DH (1988) Structure of the rabbit fast-twitch skeletal muscle Ca^{2+}-ATPase gene. J Biol Chem 263:4813–4819

Krenács T, Molnár E, Dobó E, Dux L (1989) Skeletal muscle fiber typing with sarcoplasmic reticulum Ca^{2+}-ATPase and myoglobin immunohistochemistry. Histochem J 21: 145–155

Krüger P (1952) Tetanus und Tonus der quergestreiften Skelettmuskeln der Wirbeltiere und des Menschen. Geest und Portig, Leipzig

Kucera J, Walro JM (1989) Nonuniform expression of myosin heavy chain isoforms along the length of cat intrafusal muscle fibers. Histochemistry 92:291–299

Kugelberg E (1973) Histochemical composition, contraction speed and fatiguability of rat soleus motor units. J Neurol Sci 20:177–198

Kugelberg E, Edström L (1968) Differential histochemical effects of muscle contractions on phosphorylase and glycogen in various types of fibres: relation to fatigue. J Neurol Neurosurg Psychiatry 31:415–423

Kugelberg E, Lindegren B (1979) Transmission and contraction fatigue of rat motor units in relation to succinate dehydrogenase activity of motor unit fibres. J Physiol (Lond) 288:285–300

Kugelberg E, Thornell L-E (1983) Contraction time, histochemical type, and terminal cisternae volume of rat motor units. Muscle Nerve 6:149–153

Langner BG, Pepe FA (1980) New, rapid methods for purifying α-actinin from chicken gizzard and chicken pectoral muscle. J Biol Chem 255:5429–5434

Larsson L, Ansved T (1985) Effects of long-term physical training and detraining on enzyme histochemical and functional skeletal muscle characteristics in man. Muscle Nerve 8:714–722

Laszewski-Williams B, Ruff RL, Gordon AM (1989) Influence of fiber type and muscle source on Ca^{2+} sensitivity of rat fibers. Am J Physiol 256:C420–C427

Leavis PC, Gergely J (1984) Thin filament proteins and thin filament-linked regulation of vertebrate muscle contraction. CRC Crit Rev Biochem 16:235–305

Leberer E, Pette D (1984) Lactate dehydrogenase isozymes in type I, IIA and IIB fibres of rabbit skeletal muscles. Histochemistry 80:295–298

Leberer E, Pette D (1986a) Immunochemical quantitation of sarcoplasmic reticulum Ca-ATPase, of calsequestrin and of parvalbumin in rabbit skeletal muscles of defined fiber composition. Eur J Biochem 156:489–496

Leberer E, Pette D (1986b) Neural regulation of parvalbumin expression in mammalian skeletal muscle. Biochem J 235:67–73

Leberer E, Pette D (1990) Influence of neuromuscular activity on the expression of parvalbumin in mammalian skeletal muscle. In: Pette D (ed) The dynamic state of muscle fibers. de Gruyter, Berlin, pp 497–508

Leberer E, Seedorf U, Pette D (1986) Neural control of gene expression in skeletal muscle. Ca-sequestering proteins in developing and chronically stimulated rabbit skeletal muscles. Biochem J 239:295–300

Leberer E, Klug GA, Seedorf U, Pette D (1987) Regulation of parvalbumin concentration in mammalian muscle. In: Means AR, Conn PM (eds) Cellular regulators, A: calcium- and calmodulin-binding proteins. Academic, London, pp 763–776 (Methods in enzymology, vol 139)

Leberer E, Härtner K-T, Pette D (1988) Postnatal development of Ca^{2+}-sequestration by the sarcoplasmic reticulum of fast and slow muscles in normal and dystrophic mice. Eur J Biochem 174:247–253

Leberer E, Härtner K-T, Brandl CJ, Fujii J, Tada M, MacLennan DH, Pette D (1989) Slow/cardiac sarcoplasmic reticulum Ca-ATPase and phospholamban mRNAs are expressed in chronically stimulated rabbit fast-twitch muscle. Eur J Biochem 185:51–54

Libri D, Lemonnier M, Meinnel T, Fiszman MY (1989) A single gene codes for the β subunits of smooth and skeletal muscle tropomyosin in the chicken. J Biol Chem 264:2935–2944

Locker RH, Hagyard CJ (1967) Variations in the small subunits of different myosins. Arch Biochem Biophys 122:521–522

Locker RH, Leet NG (1976) Histology of highly-stretched beef muscle. II. Further evidence on the location and nature of gap filaments. J Ultrastruct Res 55:157–172

Locker RH, Wild DJC (1986) A comparative study of high molecular weight proteins in various types of muscle across the animal kingdom. J Biochem (Tokyo) 99:1473–1484

Lompré A-M, Nadal-Ginard B, Mahdavi V (1984) Expression of the cardiac ventricular α- and β-myosin heavy chain genes in developmentally and hormonally regulated. J Biol Chem 259:6437–6446

Long L, Fabian F, Mason DT, Wikman-Coffelt J (1977) New cardiac myosin characterized from the canine atria. Biochem Biophys Res Commun 76:626–635

Lowey S (1980) An immunological approach to the isolation of myosin isoenzymes. In: Pette D (ed) Plasticity of muscle. de Gruyter, Berlin, pp 69–81

Lowey S, Risby D (1971) Light chains from fast and slow muscle myosins. Nature 278:81–85

Lowey S, Slayter HS, Weeds AG, Baker H (1969) Substructure of the myosin molecule. I. Subfragments of myosin by enzymatic degradation. J Mol Biol 42:1–29

Lowey S, Benfield P, Silberstein L, Lang LM (1979) Distribution of light chains in fast skeletal myosin. Nature 282:522–524

Lowey S, Benfield PA, LeBlanc DD, Waller GS (1983) Myosin isozymes in avian skeletal muscles. I. Sequential expression of myosin isozymes in developing chicken pectoralis muscles. J Muscle Res Cell Motil 4:695–716

Lowey S, Sartore S, Gauthier GF, Waller GS, Hobbs AW (1986) Myosin isozyme transitions in embryonic chicken pectoralis muscle. In: Molecular biology of muscle development. Liss, New York, pp 225–236

Lowry OH, Passonneau JV (1972) A flexible system of enzymatic analysis. Academic, New York

Lowry CV, Kimmey JS, Felder S, Chi MM-Y, Kaiser KK, Passonneau PN, Kirk KA, Lowry OH (1978) Enzyme patterns in single human muscle fibers. J Biol Chem 253:8269–8277

Lowry OH, Lowry CV, Chi MM-Y, Hintz CS, Felder S (1980) Enzymological heterogeneity of human muscle fibers. In: Pette D (ed) Plasticity of muscle. de Gruyter, Berlin, pp 3–18

Luff AR, Atwood HL (1971) Changes in the sarcoplasmic reticulum and transverse tubular system of fast and slow skeletal muscles of the mouse during postnatal development. J Cell Biol 51:369–383

Luginbuhl AJ, Dudley GA, Staron RS (1984) Fiber type changes in rat skeletal muscle after intense interval training. Histochemistry 81:55–58

Lutz H, Ermini M, Jenny E (1978) The size of the fibre populations in rabbit skeletal muscles as revealed by indirect immunofluorescence with anti-myosin sera. Histochemistry 57:223–235

Lutz H, Weber R, Billeter R, Jenny E (1979) Fast and slow myosin within single skeletal fibres of adult rabbits. Nature 281:142–144

Lyons GE, Haselgrove J, Kelly AM, Rubinstein NA (1983) Myosin transitions in developing fast and slow muscles of the rat hindlimb. Differentiation 25:168–175

Mabuchi K, Sréter FA (1978) Use of cryostat sections for measurement of Ca^{2+} uptake by sarcoplasmic reticulum. Anal Biochem 86:733–742

Mabuchi K, Sréter FA (1980) Actomyosin ATPase. II. Fiber typing by histochemical ATPase reaction. Muscle Nerve 3:233–239

Mabuchi K, Szvetko D, Pinter K, Sréter FA (1982) Type IIB to IIA fiber transformation in intermittently stimulated rabbit muscles. Am J Physiol 242:C373–C381

Mabuchi K, Pinter K, Mabuchi MS, Sréter F, Gergely J (1984) Characterization of rabbit masseter muscle fibers. Muscle Nerve 7:431–438

MacLennan DH, Wong PTS (1971) Isolation of a calcium sequestering protein from sarcoplasmic reticulum. Proc Natl Acad Sci USA 68:1231–1235

MacLennan DH, Campbell KP, Reitmeier RAF (1983) Calsequestrin. In: Cheung WY (ed) Calcium and cell function, vol 4. Academic, New York pp 151–173

Mahdavi V, Strehler EE, Periasamy M, Wieczorek DF, Izumo S, Nadal-Ginard B (1986) Sarcomeric myosin heavy chain gene family: organization and pattern of expression. Med Sci Sports Exerc 18:299–308

Maier A, Gambke B, Pette D (1986a) Degeneration-regeneration as a mechanism contributing to the fast to slow conversion of chronically stimulated fast-twitch rabbit muscle. Cell Tissue Res 244:635–643

Maier A, Leberer E, Pette D (1986b) Distribution of sarcoplasmic reticulum Ca-ATPase and of calsequestrin in rabbit and rat skeletal muscle fibers. Histochemistry 86:63–69

Maier A, Gorza L, Schiaffino S, Pette D (1988a) A combined histochemical and immunohistochemical study on the dynamics of fast to slow fiber transformation in chronically stimulated rabbit muscle. Cell Tissue Res 254:59–68

Maier A, Gambke B, Pette D (1988b) Immunohistochemical demonstration of embryonic myosin heavy chains in adult mammalian intrafusal fibers. Histochemistry 88:267–271

Maréchal G, Schwartz K, Beckers-Bleukx G, Ghins E (1984) Isozymes of myosin in growing and regenerating rat muscles. Eur J Biochem 138:421–428

Maréchal G, Biral D, Beckers-Bleukx G, Colson-van Schoor M (1989) Subunit composition of native myosin isoenzymes of some striated mammalian muscles. Acta Biomed Biochim 48:417–421

Margossian SS, Bhan AK, Slayter HS (1983) Role of the regulatory light chains in skeletal muscle actomyosin ATPase and in minifilament formation. J Biol Chem 258: 13359–13369

Margreth A, Angelini C, Valfre C, Salviati G (1970) Developmental patterns of LDH isozymes in fast and slow muscles of the rat. Arch Biochem Biophys 141:374–377

Margreth A, Salviati G, DiMauro S, Turati G (1972) Early biochemical consequences of denervation in fast and slow skeletal muscles and their relationship to neural control over muscle differentiation. Biochem J 126:1099–1110

Margreth A, Salviati G, Carraro U, Mussini I (1974) Specialization of the sarcoplasmic reticulum for Ca^{2+}-transport and its relationship to muscle twitch-contraction pattern. In: Drabikowski W, Strzelecka-Golaszewska H, Carafoli E (eds) Calcium binding proteins, PWN, Warsaw/Elsevier, Amsterdam, pp 519–545

Margreth A, Salviati G, Dalla Libera L, Betto R, Biral D, Salvatori S (1980a) Transition in membrane macromolecular composition and in myosin isozymes during development of fast-twitch and slow-twitch muscles. In: Pette D (ed) Plasticity of muscle. de Gruyter, Berlin, pp 193–208

Margreth A, Dalla Libera L, Salviati G, Ischia N (1980b) Spinal transection and the postnatal differentiation of slow myosin isoenzymes. Muscle Nerve 3:483–486

Martin TP, Bodine-Fowler S, Edgerton VR (1988a) Coordination of electromechanical and metabolic properties of cat soleus motor units. Am J Physiol 255:C684–C693

Martin TP, Bodine-Fowler S, Roy RR, Eldred E, Edgerton VR (1988b) Metabolic and fiber size properties of cat tibialis anterior motor units. Am J Physiol 255:C43–C50

Martonosi AN, Beeler TJ (1984) The mechanism of Ca^{2+} transport by sarcoplasmic reticulum. In: Peachy LD, Adrian RH (eds) Handbook of physiology. Am Physiol Soc, Bethesda, pp 417–485

Maruyama K, Ebashi S (1965) α-actinin, a new structural protein from striated muscle. II. Action on actin. J Biochem (Tokyo) 58:13–19

Maruyama K, Matsubara S, Natori R, Nonomura Y, Kimura S, Ohashi K, Marakami F, Handa S, Eguchi G (1977) Connectin, an elastic protein of muscle. J Biochem (Tokyo) 82:317–337

Maruyama K, Matsuno A, Higuchi H, Shimaoka S, Kimura S, Shimizu T (1989) Behaviour of connectin (titin) and nebulin in skinned muscle fibres released after extreme stretch as revealed by immunoelectron microscopy. J Muscle Res Cell Motil 10:350–359

Masaki T (1974) Immunochemical comparison of myosins from chicken cardiac, fast white, slow red and smooth muscle. J Biochem (Tokyo) 76:441–449

Masaki T, Takaiti O (1972) Purification of M-protein. J Biochem (Tokyo) 71:355–357

Masaki T, Takaiti O (1974) M-protein. J Biochem (Tokyo) 75:367–380

Masaki T, Endo M, Ebashi S (1967) Localization of 6S component of α-actinin at Z-band. J Biochem (Tokyo) 62:630–632

Mascarello F, Carpene E, Veggetti A, Rowlerson A, Jenny E (1982) The tensor tympani muscle of cat and dog contains IIM and slow-tonic fibres: an unusual combination of fibre types. J Muscle Res Cell Motil 3:363–374

Mascarello F, Veggetti A, Carpene E, Rowlerson A (1983) An immunohistochemical study of the middle ear muscles of some carnivores and primates, with special reference to the IIM and slow-tonic fibre types. J Anat 137:95–108

Matoba H, Gollnick PD (1984) Influence of ionic composition, buffering agent, and pH on the histochemical demonstration of myofibrillar actomyosin ATPase. Histochemistry 80:609–614

Matsuda G (1983) The light chains of muscle myosin: its structure, function, and evolution. Adv Biophys 16:185–218

Matsuda R, Obinata T, Shimada Y (1981) Types of troponin components during development of chicken skeletal muscle. Dev Biol 82:11–19

Matsuda R, Bandman E, Strohman RC (1983) Regional differences in the expression of myosin light chains and tropomyosin subunits during development of chicken breast muscle. Dev Biol 95:484–491

Medford RM, Nguyen HT, Destree AT, Summers E, Nadal-Ginard B (1984) A novel mechanism of alternative RNA splicing for the developmentally regulated generation of troponin T isoforms from a single gene. Cell 38:409–421

Meijer AEFH (1970) Histochemical method for the demonstration of myosin adenosine triphosphatase in muscle tissues. Histochemie 22:51–58

Mikawa T, Takeda S, Shimizu T, Kitaura T (1981) Gene expression of myofibrillar proteins in single muscle fibers of adult chicken: micro two dimensional gel electrophoretic analysis. J Biochem (Tokyo) 89:1951–1962

Minty AJ, Alonso S, Caravatti M, Buckingham E (1982) A fetal skeletal muscle actin mRNA in the mouse and its identity with cardiac actin mRNA. Cell 30:185–192

Minty A, Blau H, Kedes L (1986) Two-level regulation of cardiac actin gene transcription: muscle-specific modulating factors can accumulate before gene activation. Mol Cell Biol 6:2137–2148

Mizusawa H, Takagi A, Sugita H, Toyokura Y (1982) Coexistence of fast and slow types of myosin light chains in a single fiber of rat soleus muscle. J Biochem 91:423–425

Moore GE, Schachat FH (1985) Molecular heterogeneity of histochemical fibre types: a comparison of fast fibres. J Muscle Res Cell Motil 6:513–524

Moore GE, Briggs MM, Schachat FH (1987) Patterns of troponin T expression in mammalian fast, slow and promiscuous muscle fibres. J Muscle Res Cell Motil 8:13–22

Moss RL (1982) The effect of calcium on the maximum velocity of shortening in skinned skeletal muscle fibres of the rabbit. J Muscle Res Cell Motil 3:295–311

Moss RL, Giulian GG, Greaser ML (1982) Physiological effects accompanying the removal of myosin LC2 from skinned skeletal muscle fibers. J Biol Chem 257:8588–8591

Moss RL, Lauer MR, Giulian GG, Greaser ML (1986) Altered Ca^{2+} dependence of tension development in skinned skeletal muscle fibers following modification of troponin by partial substitution with cardiac troponin C. J Biol Chem 261:6096–6099

Moss RL, Reiser PJ, Greaser ML, Eddinger TJ (1990) Varied expression of myosin alkali light chains is associated with altered speeds of contraction in rabbit fast-twitch skeletal muscles. In: Pette D (ed) The dynamic state of muscle fibers. de Gruyter, Berlin, pp 355–368

Müntener M (1979) Variable pH dependence of the myosin-ATPase in different muscles of the rat. Histochemistry 62:299–304

Müntener M, Rowlerson AM, Berchtold MW, Heizmann MW (1987) Changes in concentration of the calcium-binding parvalbumin in cross-reinnervated rat muscles. Comparison of biochemical with physiological and histochemical parameters. J Biol Chem 262:465–469

Murakami U, Uchida K (1985) Contents of myofibrillar proteins in cardiac, skeletal, and smooth muscles. J Biochem (Tokyo) 98:187–197

Nabeshima Y, Fujii-Kuriyama Y, Muramatsu M, Ogata K (1984) Alternative transcription and two modes of splicing result in two myosin light chains from one gene. Nature 308:333–338

Nakamura A, Sréter F, Gergely J (1971) Comparative studies of light meromyosin paracrystals derived from red, white, and cardiac muscle myosins. J Cell Biol 49:883–898

Narusawa M, Fitzsimons RB, Izumo S, Nadal-Ginard B, Rubinstein NA, Kelly AM (1987) Slow myosin in developing rat skeletal muscle. J Cell Biol 104:447–459

Nemeth PM, Lowry OH (1984) Myoglobin levels in individual human skeletal muscle fibers of different types. J Histochem Cytochem 32:1211–1216

Nemeth PM, Pette D (1980) The interrelationship of two systems of fiber classification in rat EDL muscle. J Histochem Cytochem 28:193

Nemeth PM, Pette D (1981a) The limited correlation of myosin-based and metabolism-based classifications of skeletal muscle fibers. J Histochem Cytochem 29:89–90

Nemeth PM, Pette D (1981b) Succinate dehydrogenase activity in fibres classified by myosin ATPase in three hind limb muscles of rat. J Physiol (Lond) 320:73–80

Nemeth PM, Turk WR (1984) Biochemistry of rat single muscle fibres in newly assembled motor units following nerve crush. J Physiol (Lond) 355:547–555

Nemeth PM, Wilkinson RS (1990) Metabolic uniformity of the motor unit. In: Pette D (ed) The dynamic state of muscle fibers. de Gruyter, Berlin, pp 233–245

Nemeth PM, Hofer HW, Pette D (1979) Metabolic heterogeneity of muscle fibres classified by myosin ATPase. Histochemistry 63:191–201

Nemeth PM, Pette D, Vrbová G (1981) Comparison of enzyme activities among single muscle fibres within defined motor units. J Physiol (Lond) 311:489–495

Nemeth PM, Norris BJ, Solanki L, Kelly AM (1989) Metabolic specialization in fast and slow muscle fibers of the developing rat. J Neurosci 9:2336–2343

Nguyen HT, Gubits RM, Wydro RM, Nadal-Ginard B (1982) Sarcomeric myosin heavy chain is coded by a highly conserved multigene family. Proc Natl Acad Sci USA 79:5230–5234

Noegel A, Witke W, Schleicher M (1987) Calcium-sensitive non-muscle α-actinin contains EF-hand structures and highly conserved regions. FEBS Lett 221:391–396

Obinata T, Maruyama K, Sugita H, Kohama K, Ebashi S (1981) Dynamic aspects of structural proteins in vertebrate skeletal muscle. Muscle Nerve 4:456–488

Obinata T, Reinach FC, Bader DM, Masaki T, Kitani S, Fischman DA (1984) Immunochemical analysis of C-protein isoform transitions during the development of chicken skeletal muscle. Dev Biol 101:116–124

Offer G, Moos C, Starr R (1973) A new protein of the thick filaments of vertebrate skeletal myofibrils. J Mol Biol 74:653–676

Ogata T (1958a) A histochemical study of the red and white muscle fibers, I: activity of the succinoxydase system in muscle fibers. Acta Med Okayama 12:216–227

Ogata T (1958b) A histochemical study of the red and white muscle fibers. II. Activity of the cytochrome oxidase in muscle fibers. Acta Med Okayama 12:228–232

Ogata T (1958c) A histochemical study of the red and white muscle fibers. III. Activity of the diphosphopyridine nucleotide diaphorase and triphosphopyridine nucleotide diaphorase in muscle fibers. Acta Med Okayama 12:233–240

Ogata T (1988) Morphological and cytochemical features of fiber types in vertebrate skeletal muscle. CRC Crit Rev Anat Cell Biol 1:229–275

Ogata T, Mori M (1964) Histochemical study of oxidative enzymes in vertebrate muscles. J Histochem Cytochem 12:171–182

Ogata T, Yamasaki Y (1985) Scanning electron-microscopic studies on the three-dimensional structures of sarcoplasmic reticulum in the mammalian red, white and intermediate muscle fibers. Cell Tissue Res 242:461–467

Ohtsuki I, Maruyama K, Ebashi S (1986) Regulatory and cytoskeletal proteins of vertebrate skeletal muscle. Adv Protein Chem 38:1–67

Opie OH, Newsholme EA (1967) The activities of fructose 1,6-diphosphatase, phosphofructokinase and phosphoenolpyruvate carboxykinase in white muscle and red muscle. Biochem J 103:391–399

Padykula HA, Gauthier GF (1967) Morphological and cytochemical characteristics of fiber types in normal mammalian skeletal muscle. In: Milhorat AT (ed) Exploratory concepts in muscular dystrophy and related disorders, Excerpta Medica, New York pp 117–131

Padykula HA, Herman E (1955) Factors affecting the activity of adenosine triphosphatase and other phospatases as measured by histochemical techniques. J Histochem Cytochem 3:161–167

Parry DJ, Zardini D (1990) Characterization of IIx fibres in mouse muscles. In: Pette D (ed) The dynamic state of muscle fibers. de Gruyter, Berlin, pp 343–354

Pattengale PK, Holoszy JO (1967) Augmentation of skeletal muscle myoglobin by a program of treadmill running. Am J Physiol 213:783–785

Pavlath GK, Rich K, Webster SG, Blau HM (1989) Localization of muscle gene products in nuclear domains. Nature 337:570–573

Pedrosa F, Butler-Browne GS, Dhoot GK, Fischman DA, Thornell L-E (1989) Diversity in expression of myosin heavy chain isoforms and M-band proteins in rat muscle spindles. Histochemistry 92:185–194

Pepe FA, Drucker B (1975) The myosin filament. III. C-protein. J Mol Biol 99:609–617

Periasamy M, Strehler EE, Garfinkel LI, Gubits RM, Ruiz-Opazo N, Nadal-Ginard B (1984) Fast skeletal muscle myosin light chains 1 and 3 are produced from a single gene by a combined process of differential RNA transcription and splicing. J Biol Chem 259: 13595–13604

Periasamy M, Wadgaonkar R, Kumar C, Martin BJ, Siddiqui MAQ (1989) Characterization of a rat myosin alkali light chain gene expressed in ventricular and slow twitch skeletal muscles. Nucl Acids Res 17:7723–7734

Pernelle J-J, Chafey P, Lognonne J-L, Righetti PG, Bosisio AB, Wahrmann JP (1986) High-resolution two-dimensional electrophoresis of myofibrillar proteins with immobilized pH gradients. Electrophoresis 7:159–165

Pernelle J-J, Righetti PG, Wahrmann JP (1988) 2-D analysis of human skeletal muscle myosin light chains with immobilized pH gradients. J Biochem Biophys Methods 16:227–236

Perrie WT, Bumford SJ (1986) Electrophoretic separation of myosin isoenzymes. Implications for the histochemical demonstration of fibre types in biopsy specimens of human skeletal muscle. J Neurol Sci 73:89–96

Perry SV (1974) Variation in the contractile and regulatory proteins of the myofibril with muscle type. In: Milhorat AT (ed) Exploratory concepts in muscular dystrophy, Excerpta Medica, Amsterdam, pp 319–328

Perry SV (1985) Properties of the muscle proteins – a comparative approach. J Exp Biol 115:31–42

Peter JB, Barnard RJ, Edgerton VR, Gillespie CA, Stempel KE (1972) Metabolic profiles of three fiber types of skeletal muscle in guinea pigs and rabbits. Biochemistry 11:2627–2633

Pette D (1966) Mitochondrial enzyme activities. In: Tager JM, Papa S, Quagliariello E, Slater EC (eds) Regulation of metabolic processes in mitochondria. Elsevier, Amsterdam, pp 28–50 (BBA Library, vol 7)

Pette D (1981) Microphotometric measurement of initial maximum reaction rates in quantitative enzyme histochemistry in situ. Histochem J 13:319–327

Pette D (1984) Activity-induced fast to slow transitions in mammalian muscle. Med Sci Sports Exerc 16:517–528

Pette D (1985) Metabolic heterogeneity of muscle fibres. J Exp Biol 115:179–189

Pette D (1986) Regulation of phenotype expression in skeletal muscle fibers by increased contractile activity. In: Saltin B (ed) Biochemistry of exercise, vol 6. Human Kinetics Publ, Champaign, pp 3–26

Pette D (1989) Early effects of chronic low frequency stimulation on calcium-sequestering proteins and contractile properties of rabbit fast-twitch muscle. In: Rose FC, Jones R (eds) Neuromuscular stimulation. Demos, New York, pp 185–191

Pette D (1990) Dynamics of stimulation-induced fast-to-slow transitions in protein isoforms of the thick and thin filament. In: Pette D (ed) The dynamic state of muscle fibers. de Gruyter, Berlin, pp 415–428

Pette D, Bücher T (1963) Proportionskonstante Gruppen in Beziehung zur Differenzierung der Enzymaktivitätsmuster von Skelett-Muskeln des Kaninchens. Hoppe-Seyler's Z Physiol Chem 331:180–195

Pette D, Dölken G (1975) Some aspects of regulation of enzyme levels in muscle energy-supplying metabolism. In: Weber G (ed) Advances in enzyme regulation, vol 13. Pergamon, New York, pp 355–377

Pette D, Hofer HW (1980) The constant proportion enzyme group concept in the selection of reference enzymes in metabolism. In: Trends in enzyme histochemistry and cytochemistry, Ciba Foundation symposium 73 (new series), Excerpta Medica, Amsterdam, pp 231–244

Pette D, Schnez U (1977a) Myosin light chain patterns of individual fast and slow twitch fibres of rabbit muscles. Histochemistry 54:97–107

Pette D, Schnez U (1977b) Coexistence of fast and slow type myosin light chains in single muscle fibres during transformation as induced by long term stimulation. FEBS Lett 83:128-130

Pette D, Spamer C (1986) Metabolic properties of muscle fibers. Fed Proc 45:2910-2914

Pette D, Staron RS (1988) Molecular basis of the phenotypic characteristics of mammalian muscle fibres. In: Evered D, Whelan J (eds) Plasticity of the neuromuscular system. Ciba Foundation symposium 138. Wiley, Chichester, pp 22-34

Pette D, Tyler KR (1983) Response of succinate dehydrogenase activity in fibres of rabbit tibialis anterior muscle to chronic stimulation. J Physiol (Lond) 338:1-9

Pette D, Vrbová G (1985) Invited review: neural control of phenotypic expression in mammalian muscle fibers. Muscle Nerve 8:676-689

Pette D, Smith ME, Staudte HW, Vrbová G (1973) Effects of long-term electrical stimulation on some contractile and metabolic characteristics of fast rabbit muscles. Pflügers Arch 338:257-272

Pette D, Henriksson J, Emmerich M (1979) Myofibrillar protein patterns of single fibres from human muscle. FEBS Lett 103:152-155

Pierobon-Bormioli S, Sartore S, Vitadello M, Schiaffino S (1980) "Slow" myosins in vertebrate skeletal muscle. An immunofluorescence study. J Cell Biol 85:672-681

Pierobon-Bormioli S, Sartore S, Dalla Libera L, Vitadello M, Schiaffino S (1981) "Fast" isomyosins and fiber types in mammalian skeletal muscle. J Histochem Cytochem 29:1179-1188

Pierobon-Bormioli S, Betto R, Salviati G (1989) The organization of titin (connectin) and nebulin in the sarcomeres: an immunocytolocalization study. J Muscle Res Cell Motil 10:446-456

Pinter K, Mabuchi K, Sréter FA (1981) Isoenzymes of rabbit slow myosin. FEBS Lett 128:336-338

Pons F, Damadei A, Leger JJ (1987) Expression of myosin light chains during fetal development of human skeletal muscle. Biochem J 243:425-430

Potter JD (1974) The content of troponin, tropomyosin, actin, and myosin in rabbit skeletal muscle myofibrils. Arch Biochem Biophys 162:436-441

Potter JD, Gergely J (1974) Troponin, tropomyosin, and actin interactions in the Ca^{2+} regulation of muscle contraction. Biochemistry 13:2697-2703

Potter JD, Johnson JD (1982) Troponin. In: Cheung WY (ed) Calcium and cell function, vol 2. Academic, New York, pp 145-173

Price KM, Littler WA, Cummins P (1980) Human atrial and ventricular myosin light-chain subunits in the adult and during development. Biochem J 191:571-580

Prince FP, Hikida RS, Hagerman FC (1976) Human muscle fiber types in power lifters, distance runners and untrained subjects. Pflügers Arch 363:19-26

Ramirez BU, Pette D (1974) Effects of long-term electrical stimulation on sarcoplasmic reticulum of fast rabbit muscle. FEBS Lett 49:188-190

Ranvier L (1873) Propriétés et structures différentes des muscles rouges et des muscles blancs chez les lapins et chez les raies. C R Acad Sci Paris 77:1030-1034

Reichmann H, Pette D (1982) A comparative microphotometric study of succinate dehydrogenase activity levels in type I, II A and II B fibres of mammalian and human muscles. Histochemistry 74:27-41

Reichmann H, Pette D (1984) Glycerolphosphate oxidase and succinate dehydrogenase activities in II A and II B fibres of mouse and rabbit tibialis anterior muscles. Histochemistry 80:429-433

Reichmann H, Hoppeler H, Mathieu-Costello O, von Bergen F, Pette D (1985) Biochemical and ultrastructural changes of skeletal muscle mitochondria after chronic electrical stimulation in rabbits. Pflügers Arch 404:1-9

Reinach FC, Masaki T, Shafiq S, Obinata T, Fischman DA (1982) Isoforms of C-protein in adult skeletal muscle: detection with monoclonal antibodies. J Cell Biol 95:78-84

Reinach FC, Masaki T, Fischman DA (1983) Characterization of the C-protein from posterior latissimus muscle of the adult chicken: heterogeneity within a single sarcomere. J Cell Biol 96:297–300

Reiser PJ, Moss RL, Giulian GG, Greaser ML (1985a) Shortening velocity in single fibers from adult rabbit soleus muscles is correlated with myosin heavy chain composition. J Biol Chem 260:9077–9080

Reiser PJ, Moss RL, Giulian GG, Greaser ML (1985b) Shortening velocity and myosin heavy chains of developing rabbit muscle fibers. J Biol Chem 260:14403–14405

Reiser PJ, Kasper CE, Moss RL (1987a) Myosin subunits and contractile properties of single fibers from hypokinetic rat muscles. J Appl Physiol 63:2293–2300

Reiser PJ, Greaser ML, Moss RL (1987b) Tension/pCa characteristics and regulatory proteins of single fibers from chicken neonatal and adult fast and slow skeletal muscles. Biophys J 51:222a

Reiser PJ, Kasper CE, Greaser ML (1988) Functional significance of myosin transitions in single fibers of developing soleus muscle. Am J Physiol C605–C613

Robbins J, Horan T, Gulick J, Kropp K (1986) The chicken myosin heavy chain family. J Biol Chem 261:6606–6612

Robert B, Daubas P, Akimenko M-A, Cohen A, Garner I, Guenet J-L, Buckingham M (1984) A single locus in the mouse encodes both myosin light chains 1 and 3, a second locus corresponds to a related pseudogene. Cell 39:129–140

Romanul FCA (1964) Enzymes in muscle. I. Histochemical studies of enzymes in individual muscle fibers. Arch Neurol 11:355–368

Romero-Herrera AE, Nasser S, Lieska NG (1982) Heterogeneity of adult human striated muscle tropomyosin. Muscle Nerve 5:713–718

Round JM, Matthews Y, Jones DA (1980) A quick, simple and reliable histochemical method for ATPase in human muscle preparations. Histochem J 12:707–710

Rowe RWD (1973) The ultrastructure of the Z discs from white, intermediate, and red fibers of mammalian striated muscles. J Cell Biol 57:261–277

Rowlerson A, Pope P, Murray J, Whalen RB, Weeds AG (1981) A novel myosin present in cat jaw-closing muscles. J Muscle Res Cell Motil 2:415–438

Rowlerson A, Heizmann CW, Jenny E (1983) Type-specific proteins of single IIM fibres from cat muscle. Biochem Biophys Res Commun 113:519–525

Rowlerson A, Gorza L, Schiaffino S (1985) Immunohistochemical identification of spindle fibre types in mammalian muscle using type-specific antibodies to isoforms of myosin. In: Boyd IA, Gladden MH (eds) The muscle spindle. Macmillan, London, pp 29–34

Roy RK, Sréter FA, Sarkar S (1979) Changes in tropomyosin subunits and myosin light chains during development of chicken and rabbit striated muscles. Dev Biol 69:15–30

Rubinstein NA, Kelly AM (1980) The sequential appearance of fast and slow myosins during myogenesis. In: Pette D (ed) Plasticity of muscle. de Gruyter, Berlin, pp 147–159

Rubinstein N, Mabuchi K, Pepe F, Salmons S, Gergely J, Sréter F (1978) Use of type-specific antimyosins to demonstrate the transformation of individual fibers in chronically stimulated rabbit fast muscles. J Cell Biol 79:252–261

Rüegg JC (1988) Calcium in muscle activation. A comparative approach. Springer, Berlin Heidelberg New York

Ruff RL (1989) Calcium sensitivity of fast- and slow-twitch human muscle fibers. Muscle Nerve 12:32–37

Ruiz-Opazo N, Nadal-Ginard B (1987) α-Tropomyosin gene organization. Alternative splicing of duplicated isotype-specific exons accounts for the production of smooth and striated muscle isoforms. J Biol Chem 262:4755–4765

Rushbrook JI, Stracher A (1979) Comparison of adult, embryonic and dystrophic myosin heavy chains from chicken muscle by sodium dodecyl sulfate polyacrylamide gel electrophoresis and peptide mapping. Proc Natl Acad Sci USA 76:4331–4334

Rushbrook JI, Weiss C, Wadevitz AG, Stracher A (1987) Myosin isoforms in normal and dystrophic chickens. Biochemistry 26:4454–4460

Rushbrook JI, Wadewitz AG, Elzinga M, Yao T-T, Somes RG Jr (1988a) Variability in the amino acid terminus of myosin light chain 1. Biochemistry 27:8953–8958

Rushbrook JI, Weiss C, Yao T-T, Lin J (1988b) Complexity of myosin species in the avian posterior latissimus dorsi muscle. J Muscle Res Cell Motil 9:552–562

Rushbrook JI, Weiss C, Yao T-T (1990) Complexity of sarcomeric myosin species at the protein level. In: Pette D (ed) The dynamic state of muscle fibers. de Gruyter, Berlin, pp 303–314

Ryan DH, Shafiq SA (1980) A freeze-fracture study of the anterior latissimus dorsi and posterior latissimus dorsi muscles of the chicken. Anat Rec 198:147–161

Salmons S, Henriksson J (1981) The adaptive response of skeletal muscle to increased use. Muscle Nerve 4:94–105

Salmons S, Sréter FA (1976) Significance of impulse activity in the transformation of skeletal muscle type. Nature 263:30–34

Saltin B, Gollnick PD (1983) Skeletal muscle adaptability: significance for metabolism and performance. In: Peachey LD, Adrian RH, Geiger SR (eds) Handbook of physiology. sect 10: skeletal muscle. Williams and Wilkins, Baltimore Md, pp 555–631

Salviati G, Betto R, Danieli-Betto D (1982) Polymorphism of myofibrillar proteins of rabbit skeletal-muscle fibres. An electrophoretic study of single fibres. Biochem J 207:261–272

Salviati G, Betto R, Danieli-Betto D, Zeviani M (1983) Myofibrillar-protein isoforms and sarcoplasmic reticulum Ca^{2+}-transport activity of single human muscle fibres. Biochem J 224:215–225

Salviati G, Betto R, Danieli-Betto D, Biasia E, Serena M, Mini M, Scarlato G (1986) Myosin light chains and muscle pathology. Neurology 36:693–697

Samaha FJ, Guth L, Albers RW (1970) Phenotypic differences between the actomyosin ATPase of the three fiber types of mammalian skeletal muscle. Exp Neurol 26:120–125

Sarkar S, Sréter FA, Gergely J (1971) Light chains of myosins from white, red, and cardiac muscles. Proc Natl Acad Sci USA 68:946–950

Sartore S, Gorza L, Schiaffino S (1982) Fetal myosin heavy chains in regenerating muscle. Nature 298:294–296

Sartore S, Mascarello F, Rowlerson A, Gorza L, Ausoni S, Vianello M, Schiaffino S (1987) Fibre types in extraocular muscles: a new myosin isoform in the fast fibres. J Muscle Res Cell Motil 8:161–172

Schachat FH, Bronson DD, McDonald OB (1980) Two kinds of slow skeletal muscle fibers which differ in their myosin light chain complements. FEBS Lett 122:80–82

Schachat FH, Bronson DD, McDonald OB (1985a) Heterogeneity of contractile proteins. A continuum of troponin-tropomyosin expression in mammalian skeletal muscle. J Biol Chem 260:1108–1113

Schachat FH, Canine AC, Briggs MM, Reedy MC (1985b) The presence of two skeletal muscle α-actinins correlates with troponin-tropomyosin expression and Z-line width. J Cell Biol 101:1001–1008

Schachat FH, Diamond MS, Brandt PW (1987) Effect of different troponin T-tropomyosin combinations on thin filament activation. J Mol Biol 198:551–554

Schachat FH, Williams RS, Schnurr CA (1988) Coordinate changes in fast thin filament and Z-line protein expression in the early response to chronic stimulation. J Biol Chem 263:13975–13978

Schachat FH, Briggs MM, Williamson EK, McGinnis H, Diamond MS, Brandt PW (1990) Expression of fast thin filament proteins. Defining fiber archetypes in a molecular continuum. In: Pette D (ed) The dynamic state of muscle fibers. de Gruyter Berlin, pp 279–291

Schantz PG, Dhoot GK (1987) Coexistence of slow and fast isoforms of contractile and regulatory proteins in human skeletal muscle fibres induced by endurance training. Acta Physiol Scand 131:147–154

Schantz P, Billeter R, Henriksson J, Jansson E (1982) Training-induced increase in myofibrillar ATPase intermediate fibres in human skeletal muscle. Muscle Nerve 5:628–636

Schaub MC, Perry SV (1969) The relaxing protein system of striated muscle. Presolution of the troponin complex into inhibitory and calcium-sensitizing factors and their relationship to tropomyosin. Biochem J 115:993–1004

Schaub MC, Jauch A, Walzthoeny D, Wallimann T (1986) Myosin light chain functions. Biomed Biochim Acta 45:39–44

Schiaffino S, Pierobon-Bormioli S (1973) Histochemical characterization of adenosine triphosphatases in skeletal muscle fibers by selective extraction procedures. J Histochem Cytochem 21:142–145

Schiaffino S, Saggin L, Viel A, Gorza L (1985) Differentiation of fibre types in rat skeletal muscle visualized with monoclonal antimyosin antibodies. J Muscle Res Cell Motil 6:60–61

Schiaffino S, Gorza L, Dones I, Cornelio F, Sartore S (1986a) Fetal myosin immunoreactivity in human dystrophic muscle. Muscle Nerve 9:51–58

Schiaffino S, Gorza L, Sartore S, Saggin L, Carli M (1986b) Embryonic myosin heavy chain as a differentiation marker of developing human skeletal muscle and rhabdomyosarcoma. A monoclonal antibody study. Exp Cell Res 163:211–220

Schiaffino S, Saggin L, Viel A, Ausoni S, Sartore S, Gorza L (1986c) Muscle fiber types identified by monoclonal antibodies to myosin heavy chains. In: Benzi G, Packer L, Siliprandi N (eds) Biochemical aspects of physical exercise. Elsevier, Amsterdam, pp 27–34

Schiaffino S, Ausoni S, Gorza L, Saggin L, Gundersen K, Lömo T (1988a) Myosin heavy chain isoforms and velocity of shortening of type 2 skeletal muscle fibres. Acta Physiol Scand 134:575–576

Schiaffino S, Gorza L, Pitton G, Saggin L, Ausoni S, Sartore S, Lömo T (1988b) Embryonic and neonatal myosin heavy chain in denervated and paralysed rat skeletal muscle. Dev Biol 127:1–11

Schiaffino S, Gorza L, Sartore S, Saggin L, Ausoni S, Vianello M, Gundersen K, Lömo T (1989) Three myosin heavy chain isoforms in type 2 skeletal muscle fibres. J Muscle Res Cell Motil 10:197–205

Schiaffino S, Gorza L, Ausoni S, Bottinelli R, Reggiani C, Larson L, Edström L, Gundersen K, Lömo T (1990) Muscle fiber types expressing different myosin heavy chain isoforms. Their functional properties and adaptive capacity. In: Pette D (ed) The dynamic state of muscle fibers. de Gruyter, Berlin, pp 329–341

Schleicher M, André E, Hartmann H, Noegel AA (1988) Actin-binding proteins are conserved from slime molds to man. Develop Genet 9:521–530

Schmalbruch H (1971) "Rote" Muskelfasern. Z Zellforsch 119:120–146

Schmalbruch H (1979) The membrane systems in different fibre types of the triceps surae muscle of cat. Cell Tissue Res 204:187–200

Schmalbruch H (1985) Skeletal muscle. Springer, Berlin Heidelberg New York (Handbuch der mikroskopischen Anatomie des Menschen, vol 2, pt 6)

Schmitt T, Pette D (1988) Type I protein is a slow isoform of troponin T. FEBS Lett 234:83–85

Schmitt TL, Pette D (1990) Correlations between troponin T and myosin heavy chain isoforms in normal and transforming rabbit muscle fibers. In: Pette D (ed) The dynamic state of muscle fibers. de Gruyter, Berlin, pp 293–302

Schwartz RG, Rothblum KN (1981) Gene switching in myogenesis: differential expression of the actin multigene family. Biochemistry 20:4122–4129

Scott BT, Simmerman HKB, Collins JH, Nadal-Ginard B, Jones LR (1988) Complete amino acid sequence of canine cardiac calsequestrin deduced by cDNA cloning. J Biol Chem 263:8958–8964

Seedorf U, Leberer E, Kirschbaum BJ, Pette D (1986) Neural control of gene expression in skeletal muscle. Effects of chronic stimulation upon lactate dehydrogenase isozymes and citrate synthase. Biochem J 239:115–120

Seidel JC (1967) Studies on myosin from red and white skeletal muscles of the rabbit. II. Inactivation of myosin from red muscles under mild alkaline conditions. J Biol Chem 242:5623–5629

Seidel U, Bober E, Lenz S, Lohse P, Goedde HW, Grzeschik KH, Arnold HH (1988) Alkali myosin light chains in man are encoded by a multigene family that includes the adult skeletal muscle, the embryonic or atrial, and nonsarcomeric isoforms. Gene 66:135–146

Shelton GD, Gardinet GH III, Bandman E (1988) Expression of fiber type specific proteins during ontogeny of canine temporalis muscle. Muscle Nerve 11:124–132

Sher J, Cardasis C (1976) Skeletal muscle fiber types in the adult mouse. Acta Neurol Scand 54:45–56

Shima K, Tashiro K, Hibi N, Tsukada Y, Hirai H (1983) Carbonic anhydrase-III immunohistochemical localization in human skeletal muscle. Acta Neuropathol 59:237–239

Simoneau J-A, Pette D (1988) Species-specific effects of chronic nerve stimulation upon tibialis anterior muscle in mouse, rat, guinea pig, and rabbit. Pflügers Arch 412:86–92

Simoneau JA, Lortie G, Boulay MR, Marcotte M, Thibault MC, Bouchard C (1985) Human skeletal muscle fiber type alteration with high-intensity intermittent training. Eur J Appl Physiol 54:250–253

Simoneau J-A, Kaufmann M, Härtner K-T, Pette D (1989) Relations between chronic stimulation-induced changes in contractile properties and the Ca^{2+}-sequestering system of rat and rabbit fast-twitch muscles. Pflügers Arch 414:629–633

Sjögaard G, Houston ME, Nygaard E, Saltin B (1978) Subgrouping of fast twitch fibers in skeletal muscles of man: a critical appraisal. Histochemistry 58:79–87

Sjöström M, Kidman S, Henriksson-Larsén K, Ängquist K-A (1982a) Z- and M-band appearance in different histochemically defined types of human skeletal muscle fibers. J Histochem Cytochem 30:1–11

Sjöström M, Ängquist K-A, Bylund A-C, Fridén J, Gustavsson L, Schersten T (1982b) Morphometric analyses of human muscle fiber types. Muscle Nerve 5:538–553

Smillie LB, Golosinska K, Reinach FC (1988) Sequences of complete cDNAs encoding four variants of chicken skeletal muscle troponin T. J Biol Chem 263:18816–18820

Smith A, Carter N, Jeffery S (1987) Increased ATPase acid stability in type 1 fibers of rat soleus. J Histochem Cytochem 35:699–701

Snow DH, Billeter R, Jenny E (1981) Myosin types in equine skeletal muscle fibres. Res Vet Sci 30:381–382

Snow DH, Billeter R, Mascarello F, Carpene E, Rowlerson A, Jenny E (1982) No classical type IIB fibres in dog skeletal muscle, Histochemistry 75:53–65

Spamer C, Pette D (1977) Activity patterns of phosphofructokinase, glyceraldehydephosphate dehydrogenase, lactate dehydrogenase and malate dehydrogenase in microdissected fast and slow fibres from rabbit psoas and soleus muscle. Histochemistry 52:201–216

Spamer C, Pette D (1979) Activities of malate dehydrogenase, 3-hydroxyacyl-CoA dehydrogenase and fructose-1,6-diphosphatase with regard to metabolic subpopulations of fast- and slow-twitch fibres in rabbit muscles. Histochemistry 60:9–19

Spamer C, Pette D (1980) Metabolic subpopulations of rabbit skeletal muscle fibres. In: Pette D (ed) Plasticity of muscle. de Gruyter, Berlin, pp 19–30

Spurway NC (1981) Objective characterization of cells in terms of microscopical parameters: an example from muscle histochemistry. Histochem J 13:269–317

Spurway NC, Rowlerson AM (1989) Quantitative analysis of histochemical and immunohistochemical reactions in skeletal muscle fibres of Rana and Xenopus. Histochem J 21:461–474

Sréter FA (1969) Temperature, pH and seasonal dependence of Ca-uptake and ATPase activity of white and red muscle microsomes. Arch Biochem Biophys 134:25–33

Sréter FA, Gergely J (1964) Comparative studies of the Mg activated ATPase activity and Ca uptake of fractions of white and red muscle homogenates. Biochem Biophys Res Commun 16:438–443

Sréter FA, Seidel JC, Gergely J (1966) Studies on myosin from red and white skeletal muscles of the rabbit. I. Adenosine triphosphatase activity. J Biol Chem 241:5772–5776

Sréter FA, Sarkar S, Gergely J (1972) Myosin light chains of slow twitch (red) muscle. Nature New Biol 239:124–125

Sréter FA, Balint M, Gergely J (1975) Structural and functional changes of myosin during development. Comparison with adult fast, slow and cardiac myosin. Dev Biol 46:317–325

Staron RS, Pette D (1986) Correlation between myofibrillar ATPase activity and myosin heavy chain composition in rabbit muscle fibers. Histochemistry 86:19–23

Staron RS, Pette D (1987a) The multiplicity of myosin light and heavy chain combinations in histochemically typed single fibres. Rabbit soleus muscle. Biochem J 243:687–693

Staron RS, Pette D (1987b) The multiplicity of myosin light and heavy chain combinations in histochemically typed single fibres. Rabbit tibialis anterior muscle. Biochem J 243:695–699

Staron RS, Pette D (1987c) Nonuniform myosin expression along single fibers of chronically stimulated and contralateral rabbit tibialis anterior muscles. Pflügers Arch 409:67–73

Staron RS, Pette D (1990) The multiplicity of myosin light and heavy chain combinations in muscle fibers. In: Pette D (ed) The dynamic state of muscle fibers. de Gruyter, Berlin, pp 315–328

Staron RS, Hikida RS, Hagerman FC (1983) Reevaluation of human muscle fast-twitch subtypes: Evidence for a continuum. Histochemistry 78:33–39

Staron RS, Gohlsch B, Pette D (1987) Myosin polymorphism in single fibers of chronically stimulated rabbit fast-twitch muscle. Pflügers Arch 408:444–450

Starr R, Offer G (1971) Polypeptide chains of intermediate molecular weight in myosin preparations. FEBS Lett 15:40–44

Starr R, Offer G (1983) H-protein and X-protein. Two new components of the thick filament of vertebrate skeletal muscle. J Mol Biol 170:675–698

Starr R, Bennet P, Offer G (1980) X-protein and its polymer. J Muscle Res Cell Motil 1:205–206

Starr R, Almond R, Offer G (1985) Location of C-protein, H-protein and X-protein in rabbit skeletal muscle fibre types. J Muscle Res Cell Motil 6:227–256

Staudte HW, Pette D (1972) Correlation between enzymes of energy-supplying metabolism as a basic pattern of organization in muscle. Comp Biochem Physiol [B] 41:533–540

Stein JM, Padykula HA (1962) Histochemical classification of individual skeletal muscle fibers of the rat. Am J Anat 110:103–123

Steinbach JH, Schubert D, Eldridge L (1980) Changes in cat muscle contractile proteins after prolonged muscle inactivity. Exp Neurol 67:655–669

Stockdale FE, Miller JB (1987) The cellular basis of myosin heavy chain isoform expression during development of avian skeletal muscles. Dev Biol 123:1–9

Stonnington HH, Engel AG (1973) Normal and denervated muscle. A morphometric study of fine structure. Neurology 23:714–724

Stracher A (1969) Evidence for the involvement of light chains in the biological functions of myosin. Biochem Biophys Res Commun 35:519–525

Strankfeld IG, Moskalenko IE (1987) Is there any difference between actins from intact and denervated muscles? Gen Physiol Biophys 6:285–295

Strehler EE, Periasamy M, Strehler-Page MA, Nadal-Ginard B (1985) Myosin light-chain 1 and 3 gene has two structurally distinct and differentially regulated promotors evolving at different rates. Mol Cell Biol 5:3168–3182

Strohman RC, Micou-Eastwood J, Glass CA, Matsuda R (1983) Human fetal muscle and cultured myotubes derived from it contain a fetal-specific myosin light chain. Science 221:955–957

Svedenhag J, Henriksson J, Sylven C (1983) Dissociation of training effects on skeletal muscle mitochondrial enzymes and myoglobin in man. Acta Physiol Scand 117:213–218

Sweeney HL, Kushmerick MJ, Mabuchi K, Gergely J, Sréter FA (1986) Velocity of shortening and myosin isozymes in two types of rabbit fast-twitch muscle fibers. Am J Physiol 251:C431–C434

Sweeney HL, Kushmerick MJ, Mabuchi K, Sréter FA, Gergely J (1988) Myosin alkali light chain and heavy chain variations correlate with altered shortening velocity of isolated skeletal muscle fibers. J Biol Chem 263:9034–9039

Swynghedauw B (1986) Developmental and functional adaptation of contractile proteins in cardiac and skeletal muscles. Physiol Rev 66:710–771

Syrovy I (1987) Isoforms of contractile proteins. Prog Biophys Mol Biol 49:1–27

Syrovy I, Delcayre C, Swynghedauw B (1979) Comparison of ATPase activity and light subunits in myosin from left and right ventricles and atria in seven mammalian species. J Mol Cell Cardiol 11:1129–1135

Syska H, Perry SV, Trayer IP (1974) A new method of preparation of troponin I (inhibitory protein) using affinity chromatography. Evidence for three different forms of troponin I in striated muscle. FEBS Lett 40:253–257

Tada M, Inui M (1983) Regulation of calcium transport by the ATPase-phospholamban system. J Mol Cell Cardiol 15:565–575

Tada M, Kirchberger MA, Katz AM (1975) Phosporylation of a 22000-dalton component of the cardiac sarcoplasmic reticulum by adenosine 3',5'-monophosphate-dependent protein kinase. J Biol Chem 250:2640–2647

Takahashi K, Hattori A (1989) α-Actinin is a component of the Z-filament, a structural backbone of skeletal muscle Z-disks. J Biochem (Tokyo) 105:529–536

Takano-Ohmuro H, Goldfine SM, Kojima T, Obinata T, Fischman DA (1989) Size and charge heterogeneity of C-protein isoforms in avian skeletal muscle. Expression of six different isoforms in chicken muscle. J Muscle Res Cell Motil 10:369–378

Takekura H, Yoshioka T (1987) Determination of metabolic profiles on single muscle fibres of different types. J Muscle Res Cell Motil 8:342–348

Taylor LD, Bandman E (1989) Distribution of fast myosin heavy chain isoforms in thick filaments of developing chicken pectoral muscle. J Cell Biol 108:533–542

Termin A, Pette D (1990) Myosin heavy chain-based isomyosins in fast-twitch and slow-twitch muscles. Eur J Biochem (in press)

Termin A, Staron RS, Pette D (1989a) Myosin heavy chain isoforms in histochemically defined fiber types of rat muscle. Histochemistry 92:453–457

Termin A, Staron RS, Pette D (1989b) Changes in myosin heavy chain isoforms during chronic low-frequency stimulation of rat fast hindlimb muscles – a single fiber study. Eur J Biochem 186:749–754

Termin A, Staron RS, Pette D (1990) Myosin heavy chain isoforms in transforming rat muscle. In: Pette D (ed) The dynamic state of muscle fibers, de Gruyter, Berlin, pp 463–472

Thornell L-E, Carlsson E, Kugelberg E, Grove BK (1987) Myofibrillar M-band structure and composition of physiologically defined rat motor units. Am J Physiol 253:C456–C468

Thornell L-E, Carlsson E, Pedrosa F (1990) M-band structure and composition in relation to fiber types. In: Pette D (ed) The dynamic state of muscle fibers. de Gruyter, Berlin, pp 369–383

Tomanek RJ (1976) Ultrastructural differentiation of skeletal muscle fibers and their diversity. J Ultrastruct Res 55:212–227

Tomanek RJ, Asmundson CR, Cooper RR, Barnard RJ (1973) Fine structure of fast-twitch and slow-twitch guinea pig muscle fibers. J Morphol 139:47–65

Toyota N, Shimada Y (1981) Differentiation of troponin in cardiac and skeletal muscles in chicken embryos as studied by immunofluorescence microscopy. J Cell Biol 91:497–504

Trayer HR, Trayer IP (1985) Differential binding of rabbit fast muscle myosin light chain isoenzymes to regulated actin. FEBS Lett 180:170–174

Tsika RW, Herrick RE, Baldwin KM (1987a) Subunit composition of rodent isomyosins and their distribution in hindlimb skeletal muscles. J Appl Physiol 63:2101–2110

Tsika RW, Herrick RE, Baldwin KM (1987b) Time course adaptations in rat skeletal muscle isomyosins during compensatory growth and regression. J Appl Physiol 63:2111–2121

Turner DC, Wallimann T, Eppenberger HM (1973) A protein that binds specifically to the M-line of skeletal muscle is identified as the muscle form of creatine kinase. Proc Natl Acad Sci USA 70:702–705

Vaage O, Gronnerod O, Dahl H, Hermansen L (1980) Subgrouping of skeletal muscle fibres in man. Acta Physiol Scand 108:41 A

Väänänen HK, Kumpulainen T, Korhonen K (1982) Carbonic anhydrase in the type I skeletal muscle fibers of the rat. J Histochem Cytochem 30:1109–1113

Väänänen HK, Takala T, Morris DC (1986) Immunoelectron microscopic localization of carbonic anhydrase III in rat skeletal muscle. Histochemistry 86:175–179

Vandekerkhove J, Weber K (1979) The complete amino acid sequence of actins from bovine aorta, bovine heart, bovine fast skeletal muscle and rabbit slow skeletal muscle. Differentiation 14:123–133

van der Laarse WJ, Diegenbach PC, Maslam S (1984) Quantitative histochemistry of three mouse hind-limb muscles: the relationship between calcium-stimulated myofibrillar ATP-ase and succinate dehydrogenase activities. Histochem J 16:529–541

van der Laarse WJ, Maslam S, Diegenbach PC (1985) Relationship between myoglobin and succinate dehydrogenase in mouse soleus and plantaris muscle fibres. Histochem J 17:1–12

van Winkle WB, Schwartz A (1978) Morphological and biochemical correlates of skeletal muscle contractility in the cat. J Cell Physiol 97:99–120

van Winkle WB, Entman ML, Bornet EP, Schwartz A (1978) Morphological and biochemical correlates of skeletal muscle contractility in the cat. II. Physiological and biochemical studies. J Cell Physiol 97:121–136

Vetter C, Reichmann H, Pette D (1984) Microphotometric determination of enzyme activities in type-grouped fibres of reinnervated rat muscle. Histochemistry 80:347–351

Vogell W, Bishai FR, Bücher T, Klingenberg M, Pette D, Zebe E (1959) Über strukturelle und enzymatische Muster in Muskeln von Locusta migratoria. Biochem Z 332:81–117

Volpe P, Damiani E, Salviati G, Margreth A (1982) Transitions in membrane composition during postnatal development of rabbit fast muscle. J Muscle Res Cell Motil 3:213–230

Wagner PD, Giniger E (1981) Hydrolysis of ATP and reversible binding of F-actin by myosin heavy chains free of all light chains. Nature 292:560–562

Wagner PD, Weeds AG (1977) Studies on the role of myosin alkali light chains. Recombination and hybridization of light chains and heavy chains in subfragment-1 preparations. J Mol Biol 109:455–473

Waller GS, Lowey S (1985) Myosin subunit interactions. Localization of the alkali light chains. J Biol Chem 260:14368–14373

Wallimann T, Eppenberger HM (1985) Localization and function of M-line-bound creatine kinase. In: Say JW (ed) Cell and muscle motility, Plenum, New York, pp 239–285

Wallimann T, Doetschman TC, Eppenberger HM (1983) Novel staining pattern of skeletal muscle M-lines upon incubation with antibodies against MM-creatine kinase. J Cell Biol 96:1772–1779

Wang K (1984) Cytoskeletal matrix in striated muscle: the role of titin, nebulin and intermediate filaments. In: Sugi H, Pollack CH (eds) Contractile mechanisms in muscle. Plenum, New York, pp 285–306

Wang K, McClure J, Tu A (1979) Titin: major myofibrillar components of striated muscle. Proc Tatl Acad Sci USA 76:3698–3702

Wang K, Williamson CL (1980) Identification of an N2 line protein of striated muscle. Proc Natl Acad Sci USA 77:3254–3258

Wang K, Ramirez-Mitchell R (1983) A network of transverse and longitudinal intermediate filaments is associated with sarcomeres of adult vertebrate skeletal muscle. J Cell Biol 96:562–570

Wang K, Wright J (1988) Architecture of the sarcomere matrix of skeletal muscle: immunoelectron microscopic evidence that suggests a set of parallel inextensible nebulin filaments anchored at the Z line. J Cell Biol 107:2199–2212

Weeds AG (1976) Light chains from slow-twitch muscle myosin. Eur J Biochem 66:157–173

Weeds AG, Burridge K (1975) Myosin from cross-reinnervated cat muscles. Evidence for reciprocal transformation of heavy chains. FEBS Lett 57:203–208

Weeds AG, Lowey S (1971) Substructure of the myosin molecule. II. The light chains of myosin. J Mol Biol 61:701–725

Weeds AG, Hall R, Spurway NCS (1975) Characterization of myosin light chains from histochemically identified fibres of rabbit psoas muscle. FEBS Lett 49:320–324

Westwood SA, Hudlická O, Perry SV (1984) Phosphorylation in vivo of the P-light chain of myosin in rabbit fast and slow skeletal muscles. Biochem J 218:841–847

Weydert A, Barton P, Harris AJ, Pinset C, Buckingham M (1987) Developmental pattern of mouse skeletal myosin heavy chain gene transcripts in vivo and in vitro. Cell 49:121–129

Whalen RG, Sell SM (1980) Myosin from fetal hearts contains the skeletal muscle embryonic light chain. Nature 286:731–733

Whalen RG, Butler-Browne GS, Gros F (1976) Protein synthesis and actin heterogeneity in calf muscle cells in culture. Proc Natl Acad Sci USA 73:2018–2022

Whalen RG, Butler-Browne GS, Gros F (1978) Identification of a novel form of myosin light chain present in embryonic muscle tissue and cultured muscle cells. J Mol Biol 126:415–431

Whalen RG, Schwartz K, Bouveret P, Sell SM, Gros F (1979) Contractile protein isozymes in muscle development: identification of an embryonic form of myosin heavy chain. Proc Natl Acad Sci USA 76:5197–5201

Whalen RG, Sell SM, Butler-Browne GS, Schwartz K, Bouveret P, Pinset-Härström I (1981) Three myosin heavy-chain isozymes appear sequentially in rat muscle development. Nature 292:805–809

Whalen RG, Sell SM, Eriksson A, Thornell L-E (1982) Myosin subunit types in skeletal and cardiac tissues and their developmental distribution. Dev Biol 91:478–484

White MG, Snow DH (1985) Quantitative histochemistry of myosin ATPase activity after acid preincubation, and succinate dehydrogenase activity in equine skeletal muscle. Acta Histochem Cytochem 18:483–493

Wieczorek DF, Periasamy M, Butler-Browne GS, Whalen RG, Nadal-Ginard B (1985) Co-expression of multiple myosin heavy chain genes, in addition to a tissue-specific one, in extraocular musculature. J Cell Biol 101:618–629

Wieczorek DF, Smith CWJ, Nadal-Ginard B (1988) The rat α-tropomyosin gene generates a minimum of six different mRNAs coding for striated, smooth, and nonmuscle isoforms by alternative splicing. Mol Cell Biol 8:679–694

Wilkinson JM (1980) Troponin C from rabbit slow skeletal and cardiac muscle is the product of a single gene. Eur J Biochem 103:179–188

Wilkinson JM, Grand RJA (1978) The amino-acid sequence of chicken fast-skeletal-muscle troponin I. Eur J Biochem 82:493–501

Wilkinson JM, Moir AJG, Waterfield MD (1984) The expression of multiple forms of troponin T in chicken fast-skeletal muscle may result from differential splicing of a single gene. Eur J Biochem 143:47–56

Winder WW, Baldwin KM, Holloszy JO (1974) Enzymes involved in ketone utilization in different types of muscle: adaptation to exercise. Eur J Biochem 47:461–467

Winkelmann DA, Lowey S (1986) Probing myosin head structure with monoclonal antibodies. J Mol Biol 188:595–612

Wydro RN, Nguyen HT, Gubits RM, Nadal-Ginard B (1983) Characterization of sarcomeric myosin heavy chain genes. J Biol Chem 258:670–678

Yamamoto K (1984) Characterization of H-protein, a component of skeletal muscle myofibrils. J Biol Chem 259:7163–7168

Yamamoto K, Moos C (1983) The C-proteins of rabbit red, white, and cardiac muscles. J Biol Chem 258:8395–8401

Yates LD, Greaser ML (1983) Troponin subunit stoichiometry and content in rabbit skeletal muscle and myofibrils. J Biol Chem 258:5770–5774

Young OA (1982) Further studies on single fibres of bovine muscles. Biochem J 203:179–184

Young OA, Davey CL (1981) Electrophoretic analysis of proteins from single bovine muscle fibres. Biochem J 195:317–327

Zimmermann K, Starzinski-Powitz A (1989) A novel isoform of myosin alkali light chain isolated from human muscle cells. Nucl Acid Res 17:10496

Zubrzycka-Gaarn E, Korczak B, Osinska H, Sarzala MG (1982) Studies on sarcoplasmic reticulum from slow-twitch muscle. J Muscle Res Cell Motil 3:191–212

Zuurveld JGEM, Wirtz P, Loermans HMT, Veerkamp JH (1985) Postnatal growth and differentiation in three hindlimb muscles of the rat. Characterization with biochemical and enzyme-histochemical methods. Cell Tissue Res 241:183–192

Zweig SE (1981) The muscle specificity and structure of two closely related fast-twitch white muscle myosin heavy chain isozymes. J Biol Chem 256:11847–11853

Rev. Physiol. Biochem. Pharmacol., Vol. 116
© Springer-Verlag 1990

Endothelial and Neuro-Humoral Control of Coronary Blood Flow in Health and Disease

EBERHARD BASSENGE[1] and GERD HEUSCH[2]

Contents

[1] Institut für Angewandte Physiologie, Universität Freiburg, Hermann-Herder Str. 7, 7800 Freiburg, FRG
[2] Abteilung für Pathophysiologie, Zentrum für Innere Medizin, Universitätsklinikum Essen, Hufelandstr. 55, 4300 Essen, FRG

Several reviews of coronary physiology have been published in the last decade (Bassenge 1984; Berne and Rubio 1979; Feigl 1983). The purpose of the present review is to provide an update of coronary physiology and pathophysiology with particular emphasis on the changes in coronary vasomotor tone that initiate or aggravate myocardial ischemia. On the one hand, there is little new information on the mechanical determinants of coronary blood flow, and the mechanism of local metabolic regulation of coronary blood flow is still not understood in adequate detail. On the other hand, however, in the last decade much attention has been focused on the importance of the endothelial cell lining in the modulation of physiological coronary vasomotion, providing a better understanding of its interaction with humoral and cellular components in the blood stream, and of its primary alterations leading to inappropriate coronary vasomotor tone and the development of coronary atherosclerosis. In addition, the functional significance of α-adrenergic coronary vasoconstriction in experimental and clinical myocardial ischemia is presented. Finally, the current state of information about other humoral and neuronal modulations of coronary vasomotor tone is reviewed.

1 Mechanical Determinants

1.1 Perfusion Pressure

As in any perfusion circuit, myocardial blood flow is related to the available driving pressure. However, the relationship between pressure and flow is more complex in the coronary circulation than in most technical perfusion systems or in the vascular beds of other organs.

Under physiological conditions, the proximal aortic pressure is the input pressure for coronary circulation, which means that the heart generates the pressure for its own perfusion. On the other hand, the aortic pressure – in concert with cardiac dimensions – determines the left ventricular afterload and, thus, myocardial work and oxygen consumption. An increase in aortic pressure therefore leads to an increase in both the coronary perfusion pressure and the amount of coronary blood flow that is metabolically necessary. Under the pathological conditions resulting from the presence of coronary stenoses, the poststenotic coronary pressure constitutes the input pressure for myocardial perfusion. Although the poststenotic pressure still depends to some extent on aortic pressure, it is also determined by the severity of the coronary stenosis, which, in turn, depends on the stenosis morphology and the distending pressure since most stenoses are compliant, i.e., their luminal cross-section depends on the pre- and poststenotic pressure (Schwartz et al. 1980).

The site and nature of the output pressure of the coronary circulation is not clear. The output pressure is certainly not simply the coronary venous pressure at the orifice of the coronary sinus into the right atrium. When coronary perfusion pressure is systematically lowered, coronary inflow ceases at pressures substantially above right atrial pressure, i.e., at about 20–40 mmHg (Bellamy 1978, 1980; Dole and Bishop 1982; Klocke et al. 1981). Two main factors appear to contribute to the relatively high zero-flow pressure in the coronary circulation: 1) The coronary circulation has some properties of a Starling resistor, which have been termed a "vascular waterfall mechanism" (Downey and Kirk 1975). When intramyocardial pressure exceeds the pressure at the venous outflow site, it becomes the limiting outflow pressure, potentially through partial vascular collapse.

The zero-flow pressure is higher in the subendocardium than in the subepicardium (Bache and Schwartz 1982). 2) The coronary circulation has a small, but significant capacitance (Chilian et al. 1988; Mates et al. 1988; Spaan et al. 1981). Thus, coronary arterial inflow, measured in an epicardial conduit coronary artery, may cease at surprisingly high perfusion pressures but the capacitance can discharge the "stored" blood volume into the microcirculation and thus maintain myocardial perfusion. The capacitance of the coronary circulation is inversely related to coronary perfusion pressure, i.e., with increasing intraluminal distending pressure the vessels stiffen (Chilian and Marcus 1985). Conversely, with decreasing coronary vasomotor tone, coronary capacitance is enhanced (Chilian et al. 1988; Dole and Bishop 1982; Mates et al. 1988). Capacitance effects on coronary blood flow become particularly important at high pulse pressures, e.g., in aortic insufficiency (Chilian and Marcus 1985).

1.2 Extravascular Coronary Compression

There is a traditional controversy over the action of cardiac contraction on coronary blood flow. Wiggers (1954) argued that systolic myocardial contraction squeezes out the blood contained in intramural vessels, enhances coronary venous outflow, and thereby promotes subsequent arterial inflow. In contrast, Sabiston and Gregg (1957) emphasized the throttling effect of cardiac contraction on arterial inflow. In the coronary microcirculation there is a progressive shift from the arterial inflow to the venous outflow pattern, with systolic flow inhibition or reversal of flow being more pronounced in the inner myocardial layers (Tillmanns et al. 1974). The apparent contradiction between arterial inflow inhibition and venous outflow facilitation may be reconciled since the squeezing effect of cardiac contraction that enhances venous outflow is active during beat-to-beat variations in coronary blood flow and particularly important for its phasic pattern (Spaan et al. 1981). However, during steady state conditions over several cardiac cycles, myocardial contraction inhibits myocardial perfusion, and this is the clinically more relevant aspect of extravascular coronary compression when the coronary reserve is limited.

To study the effects of extravascular coronary compression on myocardial perfusion, coronary vasomotor tone is usually eliminated by maximal pharmacological dilation using adenosine, dipyridamole, or similar agents (Heusch and Yoshimoto 1983a; Raff et al. 1972a). These effects have been quantitatively related to intramyocardial tissue pressure, using a variety of techniques (Armour and Randall 1971; Johnson and Di Palma 1939; Kreuzer and Schoeppe 1963; Sabbah and Stein 1982). The concept that intramyocardial tissue pressure is the extravascular compressive force acting on the coronary vasculature is closely related to the question of what constitutes outflow pressure for the coronary circulation, as discussed above. The vascular waterfall model proposed by Downey and Kirk (1975) comprises both these aspects: the intramyocardial circulation has the characteristics of a collapsible tube compressed by intramyocardial tissue pressure, which acts as the instantaneous effective outflow pressure of the coronary circulation when it exceeds coronary venous pressure.

Other studies have not attempted to characterize the physical nature of the extravascular coronary compression but rather to relate this compressive effect of myocardial contraction to the single determinants of ventricular function. Increases in heart rate reduce the diastolic perfusion interval for coronary inflow. After elimination of coronary vasomotor tone with maximal phamacological dilation, using adenosine or dipyridamole plus carbochromene, increases in heart rate increase coronary resistance (Heusch and Yoshimoto 1983a; Raff et al. 1971a). Rises in left ventricular peak and end-diastolic pressure also augment the extravascular component of coronary resistance (Heusch and Yoshimoto 1983b; Raff et al. 1971b). Increases in myocardial contractility enhance extravascular coronary resistance even at an unchanged heart rate and left ventricular pressure (Raff et al. 1971c). Conversely, during ventricular fibrillation, i.e., in the absence of synchronized myocardial contractions and "squeeze", coronary resistance is decreased (Heusch and Yoshimoto 1983b).

Extravascular compression becomes more important with decreasing coronary perfusion pressure (Raff et al. 1972b). Thus, extravascular compression is of particular functional importance in the presence of coronary stenoses and lowered poststenotic pressures. Extravascular compression is not uniformly distributed across the left ventricular wall, decreasing from the subendocardium to the subepicardium. Thus, extravascular compression is of particular importance for the transmural distribution of myocardial blood flow (Hoffman 1987). In the presence of intact coronary autoregulatory vasomotor tone, the endocardial-epicardial gradient of extravascular compression is fully compensated by a reverse epicardial-endocardial gradient in coronary autoregulatory reserve (Bache and Cobb 1977; Klocke 1987; Rouleau et al. 1979) and vascularization (Downey et al. 1975; Wüsten et al. 1977), thus maintaining a uniform transmural myocardial blood flow distribution under physiological conditions. Transmural flow distribution remains fairly uniform both at rest and during heavy exercise (von Restorff et al. 1977), even though myocardial oxygen extraction and consumption are considerably higher in the subendocardial layers, probably due to their higher mechanical strains (Holtz et al. 1977a).

In the presence of coronary stenoses, the coronary autoregulatory reserve is compromised and the effects of extravascular coronary compression on transmural myocardial blood flow distribution become fully apparent. Any increase in heart rate (Bache and Cobb 1977), left ventricular end-diastolic pressure (Ellis and Klocke 1979; Kjekshus 1973), aortic pressure (Domenech and de la Prida 1975; Hoffman 1987), or left ventricular pressure (Buss et al. 1978) will then predominantly compromise the perfusion of the subendocardial layers. Thus, extravascular compression is of major importance for the transmurally nonuniform, preferentially subendocardial manifestation of myocardial ischemia (Bache et al. 1974; Buckberg et al. 1972). The beneficial effect of nitroglycerin on ischemic myocardial blood flow can be largely attributed to a decrease in extravascular compression (Raff and Lochner 1974).

1.3 Gregg and Gardenhose Phenomena

It is generally accepted that coronary blood flow is determined by the metabolic requirements of the myocardium which are, in turn, defined by myocardial function and oxygen consumption (see below). Several studies have suggested, however, that coronary perfusion pressure and blood flow (Gregg phenomenon; Gregg 1963), or coronary perfusion pressure alone (gardenhose phenomenon; Arnold et al. 1968), can independently determine myocardial function and consequently myocardial oxygen consumption. An increase in myocardial oxygen consumption secondary to increased myocardial perfusion implies an inverse causal relationship between function and perfusion as compared to the relation during metabolic regulation, where increases in function are supposed to cause increases in perfusion. Several potential explanations have been offered. The gardenhose hypothesis states that increasing coronary perfusion pressure distends the coronary vessels, stretching adjacent myocardial fibers, and causes a local, perivascular Frank-Starling effect (Arnold et al. 1968; Cross et al. 1961). Alternatively, increases in myocardial function and oxygen consumption secondary to increases in coronary blood flow have been attributed to the recruitment of latent capillaries (Gregg 1963) or the release of catecholamines. All observations of the Gregg or gardenhose phenomena, however, have been made in isolated hearts or in anesthetized animal preparations with an already depressed contractile performance. Therefore, the apparent improvement in function may only reflect a reversal of subendocardial ischemia when coronary perfusion pressure or blood flow were increased.

Other studies have found no evidence for a dependence of myocardial function on myocardial perfusion within the autoregulatory range (Canty

1988; Ross et al. 1963; Schulz et al. 1989). In the two recent studies, when regional myocardial function was measured with sonomicrometry and regional myocardial perfusion with tracer microspheres in anesthetized pigs (Schulz et al. 1989) or conscious dogs (Canty 1988), there was absolutely no evidence for Gregg or gardenhose phenomena within the coronary autoregulatory range.

2 Autoregulation

The term "autoregulation" is often used imprecisely. Sometimes all mechanisms, including humoral and neuronal control, which tend to maintain blood flow constant, are termed "autoregulation". More specifically, autoregulation refers to the intrinsic mechanisms which maintain blood flow constant when the perfusion pressure is varied (Dole 1987).

The coronary circulation exhibits a substantial autoregulatory capacity. Coronary blood flow is maintained relatively constant when perfusion pressure is varied between 70–130 mmHg in anesthetized dogs (Fig. 1; Mosher et al. 1964). In conscious, chronically instrumented dogs the lower

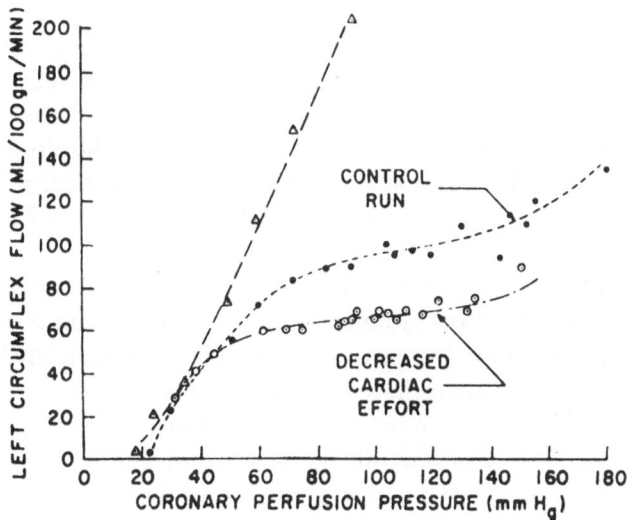

Fig. 1. Coronary autoregulation. Coronary blood flow varies in a linear fashion with instantaneous changes in coronary perfusion pressure, when starting from a perfusion pressure of 40 mmHg (*open triangles*). Autoregulation develops over time, and during steady state conditions coronary blood flow is fairly constant in a range of perfusion pressures from about 70 to 130 mmHg (*closed circles*). The whole autoregulatory curve is displaced downwards at decreased cardiac performance (*open circles*), emphasizing the coupling of coronary autoregulation to the myocardial metabolic state. From Mosher et al. (1964) by permission of the American Heart Association.

limit of autoregulation is as low as 40 mmHg (Canty 1988). When coronary perfusion pressure falls below this lower limit, myocardial blood flow becomes nearly linearly related to pressure, so that further decreases result in myocardial ischemia and contractile dysfunction (Canty 1988; Gallagher et al. 1983; Vatner 1980).

There is a transmural variation in autoregulatory capacity such that flow becomes pressure-dependent at higher perfusion pressures in the subendocardium than in the subepicardium (Canty 1988; Rouleau et al. 1979), again favoring a subendocardial manifestation of myocardial ischemia. Coronary autoregulatory adjustments involve primarily coronary arteriolar vessels, i.e., those with a diameter of less than 150 μm, but small coronary arteries can also contribute to the autoregulatory response (Chilian and Layne 1990a, Kanatsuka et al. 1990). Coronary reserve, i.e., the potential for increasing coronary blood flow above the instantaneous resting flow, is a measure of autoregulatory capacity (Klocke 1987). The human coronary circulation has a coronary reserve of four- to fivefold, i.e., at unchanged coronary perfusion pressure coronary blood flow can be increased by four- to fivefold (Marcus et al. 1981).

Whereas the existence and importance of coronary autoregulation is beyond doubt, the underlying mechanisms are less clear. (1) Increases in perfusion pressure may transiently increase myocardial perfusion and, thus, the delivery of oxygen and substrates and the removal of vasoactive catabolites. A metabolic feedback signal may then increase vascular resistance to limit the increase in coronary blood flow. This metabolic explanation of coronary autoregulation is strengthened by the close coupling of coronary autoregulatory capacity to coronary venous and, presumably, myocardial pO_2 (Dole and Nuno 1986). (2) Increases in perfusion pressure may induce a myogenic response (see below). (3) Increases in perfusion pressure may increase capillary filtration and tissue pressure and thereby limit increases in coronary blood flow. This explanation is related to extravascular compression. Increases in tissue pressure associated with autoregulatory flow adjustments have indeed been measured in response to elevated coronary perfusion pressure. However, since identical changes in tissue pressure have been measured in the absence of autoregulation, they are probably not responsible for coronary autoregulation (Driscol et al. 1964).

2.1 Myogenic Mechanisms

"Myogenic response" refers to an increase in vasomotor tone opposing a transient vascular stretch or distention secondary to an increase in perfusion pressure. Such myogenic responses were first observed by Bayliss (1902) and later more extensively studied by Folkow (1952, 1964) in various vascular preparations. Stretch, in particular dynamic stretch of a vessel, increases its electrical activity and, in consequence, also its mechanical activity (Johansson and Mellander 1975). In cerebral arteries, an intact endothelium is a prerequisite for mediating myogenic responses to increases in transmural pressure (Harder 1987). Recently, an enhanced autoregulation was observed in isolated perfused rabbit hearts after inhibition of EDRF formation (Pohl et al. 1990b). This enhancement of coronary autoregulation was attributed to enhanced myogenic responses and not to metabolic changes in coronary vasomotor tone (Pohl et al. 1990a). However, in porcine coronary arteriolar preparations no change in pressure-induced myogenic vasomotor responses was observed after mechanical endothelial denudation (Kuo et al. 1990). Thus, the role of the endothelium in modulating myogenic vasomotor responses is not completely clear at present.

In the coronary circulation evidence for myogenic mechanisms in the control of coronary blood flow is extremely difficult to obtain, since: 1) aortic pressure is not only coronary perfusion pressure but also a major determinant of both left ventricular afterload, and thus oxygen demand, and extravascular compression, and 2) even short-lasting reductions in coronary blood flow alter myocardial metabolism. Thus, it is almost impossible to distinguish myogenic from metabolic control of coronary blood flow.

Reactive hyperemia responses following very short-lasting coronary occlusions have been taken as evidence for myogenic responses (Eikens and Wilcken 1974), but even short-lasting coronary occlusions are associated with changes in myocardial metabolism (Schwartz et al. 1982). The attenuation of reactive hyperemia with the use of adenosine deaminase is approximately 30% (Dole et al. 1985; Saito et al. 1981), suggesting that myogenic mechanisms or metabolic signals other than adenosine are involved.

In general, myogenic mechanisms may be less evident in anesthetized animals and particularly in isolated preparations (McHale et al. 1987). Conversely, the best available evidence for myogenic responses in the control of coronary blood flow comes from studies in conscious dogs, in which transient changes in perfusion pressure, but not in blood flow, induced a transient increase in coronary vascular resistance (McHale et al. 1987).

2.2 Local Metabolic Regulation

Even when the heart is cut off from all external control mechanisms (nerves, hormonal and humoral factors) its ability to adjust blood flow to its metabolic requirements remains almost unaffected. This is due to a number of factors, including metabolites released from the working myocardium which act to provide an appropriate arteriolar tone. Some of these compounds may interact; for example, the dilator action of adenosine is potentiated by a fall in pH or inhibited by an increase in pH (Raberger et al. 1971). Several of the more commonly implicated metabolic regulators are discussed below.

2.2.1 Hypoxia

Hilton and Eichholz (1925) regarded tissue pO_2 of the surrounding myocardium as the decisive factor in local metabolic regulation. The hypothesis that pO_2 has a direct effect even in the absence of simultaneous major changes in metabolism recently received some support with the demonstration of a hypoxic dilator response mediated by endothelium-derived relaxing factor (EDRF) (Pohl and Busse 1989 b). In contrast, Berne et al. (1957), in experiments with controlled coronary perfusion of contracting and fibrillating dog hearts, found no coronary vasodilation when arterial oxygen content was moderately lowered. Only when coronary venous oxygen content dropped substantially (below 5.5 Vol%) was an O_2-dependent dilation observed. Therefore the decisive factor seemed to be adequate O_2 delivery, not pO_2. Dilator metabolites will only accumulate when O_2 delivery becomes insufficient (Lochner and Nasseri 1959; Guz et al. 1960). Conclusive evidence for pO_2-sensitive receptors acting independently of additional metabolites has never been presented.

2.2.2 Hypercapnia

It has been hypothesized that local metabolic regulation depends on local CO_2 concentrations and pH (Gewirtz et al. 1989; Wang and Katz 1965). Reduced myocardial perfusion leads to an accumulation of CO_2 which is highly diffusible and rapidly reaches an equilibrium between the intracellular, interstitial, and intravascular compartments. It has been suggested (Case and Greenberg 1976; Case et al. 1978) that CO_2 is involved in local metabolic regulation of myocardial blood flow. However, the effects of CO_2 are not very pronounced and are difficult to separate from hypoxia-associated metabolites such as adenosine and simultaneous changes in pH. These CO_2-mediated pH changes may alter the concentration of intracellular free Ca^{2+}, which is a decisive factor in maintaining vascular tone (Fleckenstein et al. 1975; Siegel and Schneider 1981).

2.2.3 Adenosine

In 1963, Berne observed an increase in the release of adenine nucleotide derivatives together with an increase in coronary blood flow during hypoxic perfusion of isolated, saline-perfused cat hearts and in situ blood-perfused dog hearts. Also in 1963, Gerlach et al. reported similar experiments on isolated rat hearts, in which there was an increase in myocardial tissue adenine nucleotide breakdown products with increasing duration of ischemia. Both Berne (1963) and Gerlach et al. (1963) hypothesized that the release of adenosine may serve as a metabolic feedback signal, inducing coronary vasodilation and augmenting coronary blood flow and oxygen supply in proportion to myocardial metabolic needs.

There are a number of arguments supporting the adenosine hypothesis of local metabolic coronary regulation: 1) Adenosine is, on a molar basis, one of the most powerful coronary dilators (Berne 1961, 1963) surpassed only by bradykinin (Bassenge et al. 1969, 1970). 2) The interstitial concentration of adenosine is effectively controlled. Adenosine rapidly diffuses from the cardiomyocyte cytosol into the interstitial space where coronary arterioles with their A2/Ra receptors (Burnstock 1989) are exposed to it. Interstitial adenosine may be transported back (by carrier) into the cardiomyocyte and rephosphorylated (the salvage pathway). Alternatively, it may be taken up by endothelial cells or erythrocytes and degraded to vasoinactive catabolites (for review see Berne 1980; Olsson and Bünger 1987; Schrader 1981; Schrader and Deussen 1988). 3) Under a variety of activity states with different myocardial metabolic rates, there is a good correlation between myocardial oxygen consumption, the release of adenosine, and coronary flow in vitro (Katori and Berne 1966) and in situ (Knabb et al. 1983; Miller et al. 1979).

Initially, Berne (1963) suggested that adenosine formation was coupled to myocardial oxygen tension. Later he modified his hypothesis linking adenosine formation to myocardial metabolic activity and ATP utilization (Berne 1980). However, recent experiments have revealed that adenosine formation is not triggered by changes in myocardial oxygen consumption as such, but by the relationship of myocardial oxygen demand to oxygen supply (Bardenheuer and Schrader 1983) which determines the phosphorylation potential.

More recently several problems with the adenosine hypothesis in one or the other form have become apparent: 1) Originally, Berne and associates assumed that cardiomyocytes would release AMP as the intracellular breakdown product of ATP, which would then be dephosphorylated by 5'-nucleotidase at the extracellular cell surface (Berne 1980). However, a compartmentalization of cardiac adenosine production had been previously proposed (Schrader and Gerlach 1976). Recent studies revealed that a significant fraction of the adenosine released during normoxic perfusion does not originate from cardiomyocyte breakdown products of ATP, but rather from cardiomyocyte S-adenosylhomocysteine (SAH) involved in methylation reactions (Deussen et al. 1989). Coronary vascular smooth muscle cells (Belloni et al. 1986) and, to an even greater extent, coronary endothelial cells

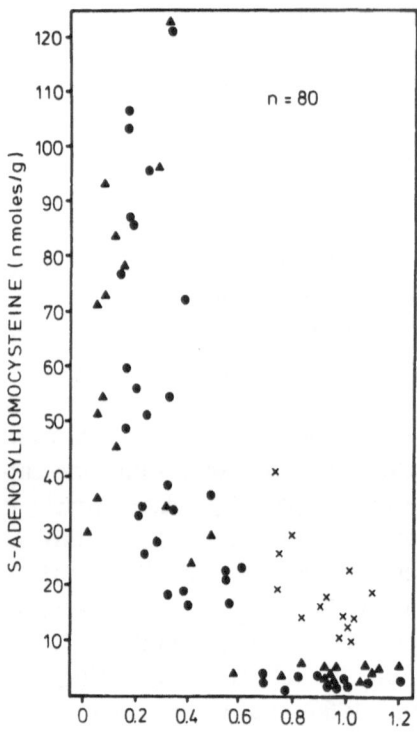

Fig. 2. Plot of S-adenosylhomocysteine which cumulatively traps the formed free intracellular adenosine versus regional myocardial blood flow in the subendocardium (*closed triangles*) and subepicardium (*closed circles*). The formation of free adenosine increases only when regional myocardial blood flow is reduced by more than 40%. *Crosses* denote samples from a border zone with normal blood flow and elevated SAH, reflecting the different spatial resolution of both measures. From Deussen et al. (1988) by permission of the American Heart Association.

contribute to adenosine release under normoxic conditions (Deussen et al. 1986; Kroll et al. 1987; Schrader and Deussen 1988). Under hypoxic conditions, however, the contribution of cardiomyocyte SAH (Deussen et al. 1989) and endothelial cells (Deussen et al. 1986) to overall adenosine release is only small. 2) During continuous infusion of norepinephrine in isolated guinea-pig hearts there is a sustained increase in myocardial oxygen consumption and coronary flow, whereas the adenosine release is only transient (DeWitt et al. 1983). Adenosine could therefore only be involved in the initial flow adaptation but not in the sustained coronary vasodilation. 3) The effective degradation of endogenous adenosine by intracoronary adenosine deaminase does not alter resting coronary blood flow in conscious dogs (Kroll and Feigl 1985). Also, adenosine deaminase does not change coronary blood flow autoregulation over a range of perfusion pressures from 40 to 100 mmHg (Dole et al. 1985). 4) Intracellular free adenosine formation is only increased when coronary blood flow is decreased by more than 40% (Fig. 2; Deussen et al. 1988). Thus, the compensatory vasodilation distal to severe coronary stenoses which reduce subendocardial blood flow by about 30% in anesthetized pigs is not attenuated by intracoronary adenosine deaminase (Gewirtz et al. 1983). The reactive hyperemia following complete coronary occlusion in anesthetized dogs is reduced by no more than 30% by adenosine deaminase, suggesting that adenosine could account for only a minor fraction of ischemic coronary vasodilation (Saito et al. 1981).

In conclusion, adenosine may make a significant contribution to initial transient coronary flow adaptations during rapid changes in cardiac performance and to hypoxic and ischemic coronary vasodilation. However, under physiological basal conditions adenosine is probably of minor importance for local metabolic coronary regulation. Due to the ample spectrum of mutual interactions, e.g., with pH or pCO_2 (Raberger et al. 1971), which experimentally can never be adequately controlled or exactly quantified, it is extremely difficult to assess the specific contribution of a single factor like adenosine within a redundant spectrum of dilator mechanisms. Recently, a common final pathway for hypoxia- and adenosine-induced coronary dilation was suggested, involving opening of ATP-sensitive K^+ channels and smooth muscle hyperpolarization (Daut et al. 1990). It remains unclear from these studies in isolated hearts, however, how smooth muscle ATP should fall rapidly enough to explain the almost instantaneous coronary dilation.

2.2.4 Hyperosmolarity and Changes in H^+, K^+, and Ca^{2+} Concentrations

The role of hyperosmolarity in mediating dilator responses is based on experimental evidence obtained primarily from rhythmically contracting skeletal muscle. Dilation of peripheral resistive vessels was attributed to increases in osmolarity (Mellander et al. 1967; for review see Mellander and Johansson 1968). In the coronary sinus effluent, only a minute hyperosmolarity was found (Scott and Radawski 1971), which probably does not explain pronounced or maximal coronary dilation.

The hyperosmolarity associated with a sudden increase in cardiac work is partially explained by increased K^+ concentration in the interstitial space. In concentrations below 15 mosmol, K^+ ions initiate coronary dilation (for review see Haddy and Scott 1975). A work load-induced release of K^+ was observed by Langer and Brady (1966) and Gerlings et al. (1969). However, this release was transient and explains only the initial, not the sustained dilation during increased cardiac work. A detailed analysis has been provided by Murray et al. (1979), who measured venous and calculated interstitial K^+ concentrations, demonstrating that half of the initial dilator response may be caused by the initial transient K^+ release.

A metabolic feedback signal through protons, independent of changes in pCO_2, has also been suggested. However, in a recent study on anesthetized pigs with coronary hypoperfusion, intracoronary infusion of sodium bicarbonate restored a normal intracellular and coronary venous pH, but did not interfere with poststenotic coronary vasodilation. This indicates that protons may not be important in metabolic coronary regulation (Gewirtz et al. 1989).

The extracellular concentration of Ca^{2+} is a determinant of coronary tone. Ca^{2+} ions cause constriction (Pitt et al. 1969a), especially when simultaneous metabolic effects are suppressed. Therefore, drugs which inhibit the transmembrane Ca^{2+} influx (calcium channel blockers) cause coronary dilation (Fleckenstein and Fleckenstein-Grün 1988). Nevertheless, changes in the interstitial Ca^{2+} concentration have never been demonstrated in physiological or pathophysiological adjustments of coronary vasomotor tone. Potentially, a pronounced hypoxia-induced release of adenosine may inhibit the transmembrane Ca^{2+} influx into the vasculature and in this way modify coronary tone (Schrader et al. 1975).

3 Endothelium

The decisive role of the endothelium in vasomotor control became clear (Fig. 3) when Furchgott and Zawadzki (1980) detected a new vasorelaxing compound, which they called endothelium-derived relaxing factor (EDRF). EDRF is probably the most important of several autacoids released from endothelial cells (e.g., prostaglandins, endothelin, platelet activating factor, etc.). Recently, it was shown that EDRF is probably the nitric oxide radical (NO·) (Ignarro et al. 1987; Ignarro 1989; Palmer et al. 1987) or a closely related compound, e.g., S-nitrosocysteine (Myers et al. 1990), though this remains controversial. EDRF directly stimulates soluble guanylate cyclase (Förstermann et al. 1986; for review see Ignarro 1989), which increases cGMP in vascular smooth muscle, causing relaxation. In platelets, EDRF inhibits activation, adhesion, and aggregation (Bassenge et al. 1989; Busse et al. 1987; Furlong et al. 1987; Pohl and Busse 1989a; Radomski et al. 1987a).

Modulation of coronary tone by the release of endothelial autacoids was first analyzed and understood in large conduit arteries such as epicardial coronary arteries, and only more recently in coronary arterioles. Arteriolar tone, however, determines coronary resistance and, therefore, myocardial perfusion under physiological conditions. Under pathophysiological conditions (Table 1), myocardial ischemia results from a critical drop in epicardial coronary conductance due to slowly progressing hyperplasia of the media, caused by atheromatosis and arteriosclerosis. Ischemia is aggravated by the loss of endothelial function, by the subsequent increased tenden-

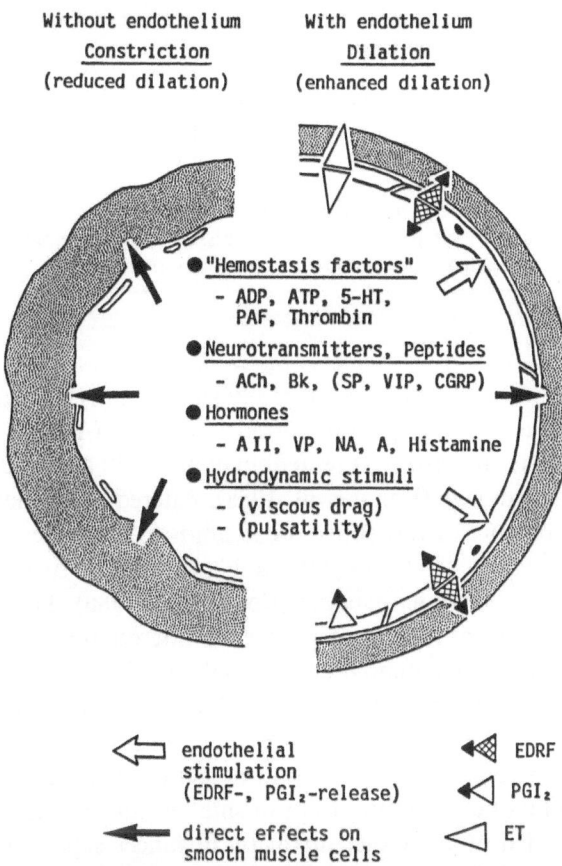

Without endothelium
Constriction
(reduced dilation)

With endothelium
Dilation
(enhanced dilation)

• "Hemostasis factors"
  - ADP, ATP, 5-HT,
    PAF, Thrombin
• Neurotransmitters, Peptides
  - ACh, Bk, (SP, VIP, CGRP)
• Hormones
  - A II, VP, NA, A, Histamine
• Hydrodynamic stimuli
  - (viscous drag)
  - (pulsatility)

⇐ endothelial
   stimulation
   (EDRF-, PGI₂-release)

◄ EDRF

← direct effects on
  smooth muscle cells

◁ PGI₂

◁ ET

Fig. 3. Endothelium-mediated control of coronary tone. In the absence of a functionally intact endothelial lining (e. g., balloon catheter-induced endothelial denudation, coronary artery disease) endogenous and exogenous compounds and factors affect vascular tone *directly*, predominantly resulting in constriction (*left hand side*). The intact endothelium has a decisive modulatory role for the various stimulatory compounds and factors (*right hand side*). Therefore, the combined net response is usually dilation or attenuated constriction. The pathophysiological significance of endothelium-dependent, thrombin-induced endothelin release and its inhibition by EDRF or nitrovasodilators is not clear yet (Boulanger and Lüscher 1990). Prostacyclin release is predominant towards the luminal side, whereas EDRF/NO·release is similar on both the luminal and the abluminal side. Substances/factors in *parentheses* only elicit vasodilation. *5-HT*, 5-hydroxytryptamine (serotonin); *PAF*, platelet activating factor; *ACh*, acetylcholine; *BK*, bradykinin; *SP*, substance P; *VIP*, vasoactive intestinal polypeptide; *CGRP*, calcitonin gene-related peptide; *AII*, angiotensin II; *VP*, vasopressin; *NA*, noradrenaline; *A*, adenosine; *ET*, endothelin.

cy for platelet activation, adhesion and aggregation, and finally, by the initiation of coronary thrombosis.

Whereas pathological changes of coronary vascular morphology and function are predominantly observed in large conduit arteries, the arterio-

Table 1. Potential involvement of impaired endothelial function and integrity in pathophysiological processes

1. Inappropriate modulation of coronary vasomotor tone
2. Enhanced platelet activation and coagulation
3. Development of atherosclerosis
4. Hypercholesterolemia
5. Acute and chronic hypertension
6. Impaired sympathetic neurotransmission and vasomotor control
7. Augmented venous tone and potentially preload

lar resistance vessels are rarely significantly affected (e.g., in diabetes). In diabetes, there is an impaired response to EDRF both in large epicardial arteries in vitro (Gebremedhin et al. 1988) and in overall coronary conductance in vivo (Koltai et al. 1988). Altered α-adrenergic mechanisms and an imbalance of vasoactive prostanoids (i.e., augmented thromboxane production) may also be involved and contribute to the high incidence of myocardial ischemia in diabetics (Koltai et al. 1988). The regulation of large coronary artery tone is of particular interest under clinical conditions in the presence of impaired endothelial function and inappropriate autacoid release. Therefore, this section will deal mainly with the control of large coronary arteries. Furthermore, these epicardial vessels are easier to analyze, as there is substantially less interaction with the usually predominant local metabolic regulation than in smaller, intramural vessels. This is easily explained by the larger diffusion distances separating the working and metabolizing parenchyma (cardiac muscle) from the epicardial feed vessels.

3.1 The NO· Radical as a Biological Signal and Modulator of Autonomic Neurotransmission

NO· is formed by a variety of cells with different biological actions (for review see Moncada et al. 1989), and is released from macrophages and leukocytes (McCall et al. 1989; Hibbs et al. 1987; Wright et al. 1989), exerting bacteriostatic, cytotoxic, and microbicidal actions. In neurons of the forebrain and cerebellum, NO· may be a modulator of neurotransmission (Garthwaite et al. 1988; Knowles et al. 1989). NO· probably plays a role in autonomic neurotransmission as well (Greenberg et al. 1989) and NO· may be the inhibitory non-cholinergic, non-adrenergic neurotransmitter or coreleased factor in various vascular beds (Bult et al. 1990) and other smooth muscle preparations (Gibson et al. 1990). Furthermore, NO· may reduce sympathetically induced vasoconstriction by a central mechanism,

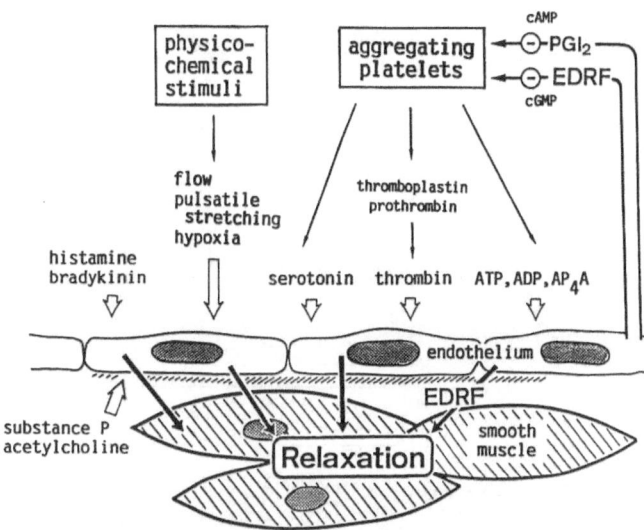

Fig. 4. Bidirectional autacoid release (mainly PGI_2 and EDRF/NO·) upon endothelial stimulation with various factors and compounds. Luminal release (*top*) stimulates soluble guanylate cyclase (sGC) in platelets (by EDRF) and adenylate cyclase (by PGI_2), eliciting inhibition of platelet adhesion and aggregation which is most effective (due to diffusion gradient) at the endothelial surface. Subthreshold concentrations of PGI_2 and EDRF potentiate each other. Abluminal release (predominantly EDRF) stimulates sGC in the vasculature.

since the inhibitor of NO·-production L-NMMA, intracisternally applied, results in a substantial enhancement of sympathetic nerve activity (Sakuma et al. 1990).

In endothelial cells, distinct luminal and abluminal EDRF/NO· production and release can be observed (Bassenge and Pohl 1986; Bassenge et al. 1987; Busse et al. 1985) which serve two purposes. Abluminally released, EDRF/NO· relaxes vascular smooth muscle. In addition, EDRF/NO· may inhibit norepinephrine release from sympathetic nerve endings (Cohen and Weisbrod 1988; Greenberg et al. 1989; Tesfamariam and Cohen 1988; Tesfamariam et al. 1987), thereby decreasing neuronally mediated coronary constrictor tone. Luminally released, EDRF/NO· inhibits platelet activation and aggregation (Fig. 4; Bassenge et al. 1989; Busse et al. 1987; Furlong et al. 1987; Pohl and Busse 1989a; Radomski et al. 1987a, b and c).

NO· is formed by a not yet identified Ca^{2+}/calmodulin- and NADPH-dependent enzyme, using L-arginine as a substrate or precursor, as indicated in Fig. 5 (Marletta et al. 1988; Moncada et al. 1989; Mülsch et al. 1989). This NO· formation can be competitively inhibited by the biologically inert analogue L-N^G-mono-methyl-arginine (L-NMMA). L-NMMA, in turn,

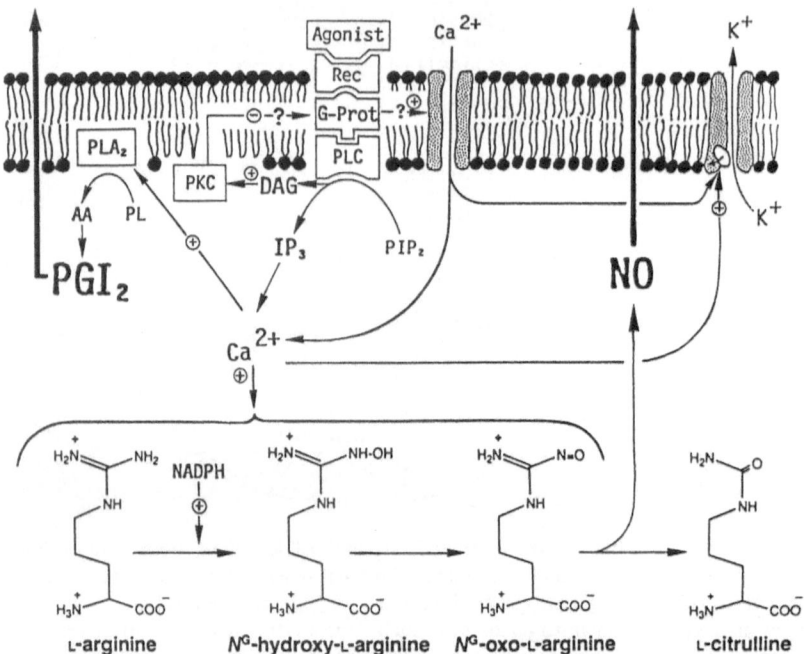

Fig. 5. Signal transduction and autacoid release from endothelial cells upon agonist (e. g., bradykinin, acetylcholine, serotonin etc.)-induced stimulation. Upon binding of an agonist to its specific membrane receptor the activity of phospholipase C (*PLC*) is changed via a coupling GTP-binding protein (*G-Prot*). PLC stimulation results in the formation of inositol 1,4,5-trisphosphate (IP_3) and diacylglycerol (*DAG*) from phosphatidylinositol diphosphate (PIP_2). IP_3 mobilizes Ca^{2+} from internal stores and increases intracellular free calcium concentration $[Ca^{2+}]_i$. These increases in $[Ca^{2+}]_i$ stimulate 1) phospholipase A_2 and hence release of arachidonic acid (AA) and synthesis of PGI_2, and 2) Ca^{2+}-dependent K^+ channels resulting in hyperpolarization and enhanced transmembrane Ca^{2+} influx (Lückhoff and Busse 1990).

Bottom, increased intracellular Ca^{2+} probably induces, in a calmodulin- and NADPH-dependent step (Busse and Mülsch 1990), an increased formation and release of EDRF/NO· from the precursor L-arginine, while the continuous basal EDRF release is probably caused by the resting intracellular free Ca^{2+} and independent of agonist-induced increases in $[Ca^{2+}]_i$ (Mülsch et al. 1989). A nitrogen atom, N^G, from L-arginine is oxidized in several steps to ultimately form L-citrulline and the radical NO·. A feed-back control mechanism adjusting the receptor-mediated endothelial autacoid release may be presented by the DAG-induced stimulation of protein kinase C (*PKC*) acting back on the modulatory G-Prot, thereby suppressing coupling from receptor to PLC.

can be replaced competitively by augmented concentrations of the biologically active stereo-enantiomer L-arginine, but not by the biologically inert D-arginine (Marletta et al. 1988; Moncada et al. 1989; Mülsch and Busse 1990; Rees et al. 1989).

3.2 EDRF/NO$\cdot$-Mediated Dilator Effects

There is a continuous, basal EDRF release from endothelial cells of the coronary system (Kelm and Schrader 1988; Stewart et al. 1987a,b). This basal release is maintained by a number of circulating agonists such as nor-epinephrine, bradykinin, and thrombin. The most important stimulus for EDRF release is probably the laminar shear stress, or viscous drag, of the circulating blood (Holtz et al. 1983; Pohl et al. 1986a; Rubanyi et al. 1986) and the pulsatile stretching of the endothelial lining (Pohl et al. 1986b). Shear stress-dependent Ca^{2+} and K^+ channels have been described recently (Lansman et al. 1987; Olesen et al. 1988). Activation of Ca^{2+} channels resulting in an increased intracellular concentration of Ca^{2+} ($[Ca^{2+}]_i$) is similar to a receptor-triggered Ca^{2+} transient in excitable cells, but less pronounced. This rise in $[Ca^{2+}]_i$ is a prerequisite for EDRF formation.

In beating, saline-perfused Langendorff-heart preparations, the basal release of EDRF can be antagonized by the addition of hemoglobin or suppressed by the addition of monomethyl-arginine to the perfusate. Hemoglobin binds and inactivates EDRF, resulting in a rise in coronary tone (Stewart et al. 1987a, b) and a decrease in myocardial perfusion. In comparable experiments in perfused skeletal muscle, a corresponding reduction of blood flow and oxygen consumption was demonstrated following gossypol-induced EDRF suppression (Pohl and Busse 1988). Recently these findings were extended by the observation of reduced coronary conductance in beating heart preparations with impaired endothelial function, obtained by the addition of the biologically inert L-arginine analogue, L-NMMA. L-NMMA addition resulted in the reversible suppression of endothelial NO$\cdot$ production and also a decrease in overall coronary conductance (Amezcua et al. 1989), which was caused neither by a reduction in cardiac work nor, presumably, by oxygen consumption. These findings have been confirmed in chronically instrumented, conscious dog preparations (Chu et al. 1989a, b).

Using the NO$\cdot$ hemoglobin (photometric) method to detect NO$\cdot$ in the effluent of beating heart preparations, it was possible to directly and continuously monitor NO$\cdot$ release and demonstrate the changes induced by various stimulator compounds, such as bradykinin (Kelm and Schrader 1988).

Recently, a substantial increase in plasma nitrate levels, derived from endothelial NO$\cdot$ production and release, was demonstrated by electron paramagnetic resonance spectrometry during reactive hyperemia following brief periods of ischemia (Wennmalm et al. 1990).

3.3 Flow- and Shear-Induced Dilator Responses

3.3.1 Activation of K^+ Channels with Hyperpolarization

When cultured endothelial monolayers were exposed to laminar shear forc-
es, a shear stress-dependent hyperpolarization, probably due to opening of
K^+ channels, of the membrane potential was recorded (Olesen et al.
1988). Accordingly, charybdotoxin which blocks calcium-activated K^+
channels suppresses flow-induced dilations whereas acetylcholine-induced
dilations are unaffected (Cooke et al. 1990). Although this hyperpolarization theoretically could spread into the adjacent vasculature and elicit
vasorelaxation, it is more likely a trigger in the process of EDRF formation
within the endothelial cells.

3.3.2 Endothelial Hyperpolarization Favors NO· Production

Several agonists like bradykinin induce endothelial hyperpolarization,
which is augmented and prolonged by the addition of an activator of K^+
channels (e.g., BRL 34915; Lückhoff and Busse 1990). This hyperpolarization is paralleled by an increased endothelial EDRF/NO· production and
release, whereas depolarization with extracellular K^+ inhibits EDRF release. The dependence of EDRF/NO· production on membrane hyperpolarization can be explained by a receptor (bradykinin)-mediated mobilization of Ca^{2+} from intracellular stores, activating K^+ channels and initiating hyperpolarization. This hyperpolarization provides an increased
driving force for transmembrane Ca^{2+} influx. NO· production from the
guanidino-terminal of L-arginine is enhanced due to facilitation of a
Ca^{2+}-dependent enzymatic step.

3.3.3 Shear Stress Elicits Increase in Intracellular Ca^{2+}

Exposure of endothelial monolayers to gradually increasing laminar shear
stress results in similarly progressive increases in cytosolic Ca^{2+} concentration (Ando et al. 1989; Dull and Davies 1989). Using the Fura-2 method,
steep (transient-like) increases in Ca^{2+} were recorded upon each stepwise
increase in shear stress, with a subsequent, rapid decline in Ca^{2+}.

3.3.4 Increase in Intracellular Ca^{2+} as a Prerequisite for
"Stimulated" EDRF Release

NO· is formed in the cytosol of endothelial cells, by a Ca^{2+}/calmodulin-
and NADPH-dependent oxidative conversion of the guanidino-terminal of

L-arginine supplied as a precursor in the cytosol. Thus, a receptor-dependent stimulation of EDRF/NO· release may be initiated. The existence of a second Ca^{2+}-independent NO· formation has recently been demonstrated (Mülsch et al. 1989) which probably accounts for the continuous basal, nonstimulated EDRF/NO· release. Depriving cells of L-arginine, or competitive replacement by the biologically inert analogues such as L-mono-methyl-arginine (L-NMMA), L-nitro-arginine or L-amino-arginine, results in the cessation of endothelial NO· production and, thus, enhanced vasoconstrictor tone. This can be reversed by the addition of L-arginine, but not D-arginine (Mülsch and Busse 1990; Rees et al. 1989).

3.4 Basal and Flow-Stimulated EDRF Release
in Large Coronary Arteries

A small, basal and continuous ("nonstimulated") release of EDRF, both from cultured (Busse et al. 1985; Cocks et al. 1985) and from genuine endothelial cells (Hartmann et al. 1987), has been demonstrated (for review see Bassenge and Busse 1988; Griffith et al. 1988). Under physiological conditions, this basal release cannot be exactly distinguished from a flow- and pulsatility-induced release, which depends on the functional integrity of the endothelial lining (Holtz et al. 1983; Inoue et al. 1988; Pohl et al. 1986a, b). These dilator effects, exerted by physical and hydrodynamic factors and superimposed on resting tone, can substantially modify and enhance the continuous basal EDRF release and its dilator action. Flow-induced shear stress, acting upon the luminal endothelial surface as viscous drag, is probably the most important stimulus in the endothelium-dependent control of vascular caliber (Holtz et al. 1983; Melkumyants et al. 1987, 1989; Smiesko et al. 1985). Such flow-dependent control seems to be of particular importance in the coronary circulation (Fig. 6; Holtz et al. 1984) because of its large moment to moment variability and because of its extremely pulsatile nature of its flow patterns – usually a distinct flow reversal in systole (for review see Bassenge 1984, 1989; Bassenge and Pohl 1986). Indeed, the pulsatile stretching of the endothelial cell lining has been demonstrated to be another important stimulus for EDRF release (Pohl et al. 1986b) and PGI_2 release (Quadt et al. 1982), in addition to the viscous drag acting upon an intact endothelial lining (Holtz et al. 1983; Inoue et al. 1988; Pohl et al. 1986a).

Acute and chronic changes in flow and viscous drag modify EDRF release from endothelial cells, thereby constituting a rapidly responding feedback mechanism for the control of vasomotor tone both in large epicardial arteries (Holtz et al. 1983) and in the coronary resistance vessels (Pohl et

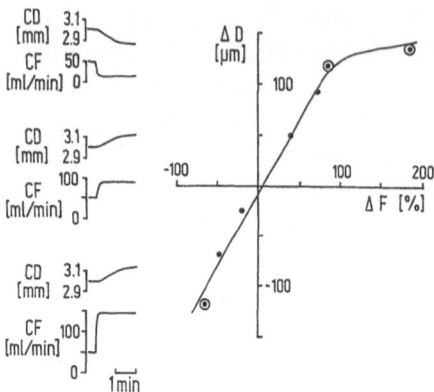

Fig. 6. Flow- or shear stress-mediated regulation of coronary diameters. When coronary flow is experimentally changed up to 200% (*ΔF, abscissa*), an endothelium-mediated dilator response (*ΔD, ordinate*) is initiated. Conversely, flow reductions result in diameter reductions. The *left hand panel* depicts some examples for simultaneous coronary diameter (*CD*) and flow (*CF*) tracings used for the *right hand* regression curve. This moment to moment adjustment of vessel diameter is probably initiated by a shear stress-induced Ca^{2+} signal in endothelial cells, resulting in an increased autacoid production, namely NO· and PGI_2 (for details see text).

al. 1990 a, b). Furthermore, abluminally released EDRF may inhibit nor-epinephrine release from nerve endings, thereby reducing sympathetic con-strictor tone (Tesfamariam and Cohen 1988). During prolonged or chronic changes, additional adaptive growth processes or caliber regressions in peripheral arteries have been reported (Kamiya and Togawa 1980; Langille and O'Donnell 1986; for review see Bassenge and Pohl 1986). These ad-justments serve to normalize the shear stress acting upon the endothelial lining, keeping it within an optimal physiological range, as predicted in an interesting analysis by Rodbard (1975). This acts to prevent chronic endo-thelial damage, functional impairment, and necrosis. Furthermore, the cal-iber adjustments are important for the spatial distribution of tissue perfu-sion and for the optimization of pressure gradients within the vascular net-work, especially when perfusion rates are suddenly altered (Griffith et al. 1988). For such caliber adjustments endothelial cells must act as "flow sen-sors" and must transduce these hydrodynamically evoked signals (acting as tangential laminar stress upon the cell membrane) into an intracellular sig-nal event, initiating a transduction cascade which finally results in an al-tered EDRF/NO· production and release.

3.5 Pathophysiological Alterations in EDRF/NO·-Mediated Regulation of Coronary Vasomotor Tone

3.5.1 Effects of Hypercholesterolemia

The equilibrium between vasoconstrictor and vasodilator mechanisms for the maintenance of adequate coronary tone is disturbed in the presence of hypercholesterolemia, atheromatosis, and arteriosclerosis favoring inappropriate coronary vasoconstriction. This has been demonstrated in in vitro experiments, animal models, and a number of recent clinical studies. Inappropriate vasomotor control is probably related to loss of endothelium-dependent vasodilator function.

3.5.1.1 Scavenger Effect on EDRF

In hypercholesterolemia, elevated levels of low density lipoproteins can scavenge abluminally released EDRF as it diffuses from endothelial cells to the adjacent vasculature, accelerating its inactivation (Galle et al. 1990b; Jacobs et al. 1990). This is similar to the action of hemoglobin as an inactivator of EDRF/NO· (through formation of NO-Hb) in saline perfused organs. Under clinical conditions, this scavenger effect has not yet been analyzed, but may be of importance in regions of vascular fatty streaks or atheromatous plaques.

3.5.1.2 Diminished EDRF/NO· Release

There is reduced EDRF/NO· release from endothelial cells during hypercholesterolemia, atheromatosis and arteriosclerosis (Verbeuren et al. 1986). Part of this reduction is explained by functional endothelial impairment and cell necrosis. However, the reason for reduced EDRF/NO· release from morphologically intact endothelial cells is unknown.

3.5.1.3 Effects of Low-Density and Oxidized Lipoproteins

The effects of low-density lipoproteins (LDL) on smooth muscle tone of in vitro vascular preparations are probably mediated by oxidatively modified LDL, since LDL undergoes rapid oxidation to form lipid peroxides and oxidized LDL (ox-LDL) (Galle et al. 1990a; Simon et al. 1989). Increased plasma LDL levels augment the activation of macrophages (Gerrity et al. 1979), favoring increased formation (Steinberg et al. 1989), storage, and deposition of ox-LDL in the vascular wall (Ylä-Herttuala et al. 1989). Endothelial cells themselves can oxidize LDL (Steinbrecher et al. 1984), and ox-LDL can be demonstrated in the vascular wall during atheromatosis and atherosclerosis (Daugherty et al. 1988; Ylä-Herttuala et al. 1989). In in vitro experiments with perfused vessel segments, ox-LDL causes potentiation of agonist-induced constrictions (e. g., by catecholamines, serotonin, angiotensin II etc.). One study revealed a dependence on the endothelium for this effect (Simon et al. 1990). However, in another study using a different species this effect is even augmented in the absence of an endothelial lining. Ox-LDL-induced vasoconstriction is inhibited in the presence of Ca^{2+} antagonists (Galle et al. 1990a). A dependence on extracellular Ca^{2+} for the potentiation of adrenergic stimulation is observed in arteries perfused with cholesterol-phospholipid-containing media (Broderick et al. 1989). Finally, norepinephrine causes net coronary constriction in hypercholesterolemic, conscious dogs rather than a predominant metabolical dilation as in normal controls (Rosendorff et al. 1981). The increased cardiovascular reactivity to sympathetic stimulation with elevated plasma lipoproteins, which may also involve increased platelet activation, was recently reviewed by Howes and Krum (1989).

Vessel segments from cholesterol-fed and atheromatotic animals show a diminished sensitivity to acetylcholine (receptor-dependent) stimulation (Freiman et al. 1986), and to EDRF/NO· superfusion (Verbeuren et al. 1986, 1990). The mechanism responsible for this desensitization has not been identified yet, but subintimal thickening, as an additional diffusion barrier to EDRF/NO·, can be excluded (Harrison et al. 1987a).

3.5.1.4 Potential Regression of Inadequate Vasomotor Responses with Therapy of Hypercholesterolemia

Studies in atherosclerotic arteries have demonstrated that endothelium-mediated dilations resulting from some, but not all, agonists are impaired. This suggests a rather selective atherosclerosis-induced deficiency of endothelial cell function. Interestingly, the impaired endothelium-dependent dilator responses completely reappear after feeding a regression diet for 3 months, when normal cholesterol levels are reestablished (Heistad et al. 1987), even though the pronounced intimal thickening persists (Harrison et al. 1987b). Partial restoration of endothelium-dependent coronary dilation was recently demonstrated angiographically in patients with coronary artery disease (CAD patients), who were placed on a dietary program to change their lipoprotein profile (Vekshtein et al. 1989). Endothelial dysfunction in the regenerated state or following hypercholesterolemia may be caused by a persistent impairment of Gi protein function (Shimokawa et al. 1989a). There still seems to be insufficient and inconsistent data on the clinically important questions of how long functional restoration takes, the extent to which impaired endothelium-dependent dilator function can be reestablished, and which factors may be favorable for such functional restoration.

3.5.2 Impaired Receptor-Mediated, Endothelium-Dependent Vasomotor Control

Femoral, aortic, and coronary vascular strips from hypercholesterolemic rabbits fed on a high-cholesterol diet for 2 months display a supersensitivity to various vasoconstrictor agonists, especially to serotonin, which may contribute to vascular dysfunction in vivo (Henry and Yokoyama 1980). The modulatory role of a functional endothelial lining in maintaining appropriate vascular tone was not known at the time these experiments were conducted, and therefore was not tested.

Later is was shown that hypercholesterolemia-, atheromatosis-, and arteriosclerosis-induced potentiation of vasoconstrictor stimuli can be explained by a loss of endothelium-dependent vasodilation.

This has been associated with an unmasking of $5HT_2$ receptors (Vrints et al. 1990) or an upregulation of H_1 receptors in human coronary arteries (Ginsburg et al. 1981; Ginsburg 1984). Loss of endothelium-dependent relaxation, especially in response to cholinergic stimulation, was demonstrated in resistance arteries of perfused hindlimbs of hypercholesterolemic rabbits (Bossaller et al. 1986). This defect occurred prior to a more generalized impairment of endothelium-dependent vasodilation, since the Ca^{2+} ionophore A 23187-induced endothelium-dependent dilation was still intact (Bossaller et al. 1987). Osborne et al. (1989a, b) found, along with the impaired endothelium-dependent relaxation in peripheral and coronary arteries of cholesterol-fed rabbits, a greater susceptibility to myocardial ischemia during experimental coronary stenosis. This could be counteracted by the lipid lowering agent lovastatin. In excised atherosclerotic coronary arteries from human transplants, there was a substantial reduction of acetylcholine-induced, endothelium-dependent relaxation in the presence of a more (Bossaller et al. 1987) or less well preserved EDRF release induced by the Ca^{2+} ionophore A 23187 (Förstermann et al. 1988).

In patients undergoing angiographic studies, the cold pressor test elicited constrictions of atherosclerotic coronary segments, whereas dilator responses were observed in normal segments (Nabel et al. 1988a, b). In isolated ring segments of coronary arteries from hypercholesterolemic pigs, which were not yet affected by fatty streaks, intimal proliferative changes, or atherosclerosis, serotonin induced constrictions rather than dilations. The A 23187-induced dilator responses were not diminished by hypercholesterolemia, nor were the dilator responses elicited by sodium nitroprusside or papaverine (Cohen et al. 1988). Comparable data were obtained in atheromatotic peripheral arteries of monkeys after diet-induced hypercholesterolemia (Heistad et al. 1984). Endothelium-dependent dilations to acetylcholine and thrombin were likewise inhibited, while those to

A 23187 were hardly affected (Freiman et al. 1986). Thus, an abnormal function of endothelial receptors in the coronary artery preceeds the onset of histologically visible coronary ateroslerosis. Platelet-related compounds such as ADP, 5-HT, and TxA_2 elicit abnormal endothelium-dependent vasomotion in peripheral arteries of hypercholesterolemic monkeys, when only slight intimal thickening is present (Lopez et al. 1989).

3.5.3 Endothelial Dysfunction in CAD Patients

Responses to endothelium-dependent dilators such as acetylcholine (Ludmer et al. 1986; Zeiher et al. 1989a), bradykinin (Rafflenbeul et al. 1988) or substance P (Crossman et al. 1989) can be used as an index of functional integrity for appropriate autacoid release from the endothelial lining of the coronary bed. In the absence of CAD, an intracoronary injection of acetylcholine results in an endothelium-mediated, angiographically documented, dose-dependent dilator response of the large coronary arteries and of the arterioles. The arteriolar response results in an increased myocardial perfusion, which in turn adds to the large artery dilator response through the shear stress-induced augmentation of EDRF/NO· release from the proximal segments. In the presence of CAD, however, when endothelium is absent or functionally impaired, constriction due to stimulation of muscarinic smooth muscle receptors results. Consequently, in CAD patients and patients with heart transplantation-induced endothelial damage (Nellessen et al. 1988), epicardial coronary artery constriction was observed instead of the normal dilation in response to acetylcholine (Hodgson and Marshall 1989; Ludmer et al. 1986; Werns et al. 1989; Zeiher et al. 1989a). The flow-dependent dilator response was substantially reduced with functional impairment of endothelial cells in CAD patients (Drexler et al. 1989; Nabel et al. 1988a, b; Vita et al. 1990). It may still be present to some extent even when local, acetylcholine-induced, direct constrictor responses can be demonstrated (Zeiher et al. 1989a). Modification of the plasma lipoprotein profile by various programs resulted in improvement or reappearance of endothelium- and flow-dependent increases in large coronary artery calibers (Vekshtein et al. 1989).

3.5.4 Hypertension and EDRF

Both acute and chronic hypertension are associated with endothelial dysfunction and impaired endothelium-mediated vascular control. Whether this impaired function is the cause for or the consequence of hypertension has not yet been determined. A number of reports demonstrate loss of or changes in endothelium-mediated dilator functions in peripheral arteries with chronic hypertension (both in vivo in man and in vitro from test animals, Linder et al. 1990). These dilator effects can be reestablished with antihypertensive therapy (Lüscher and Vanhoutte 1986; Lüscher et al. 1987). In one specific rabbit model of chronic hypertension, however, impairment of dilator function could not be detected (Wright and Angus 1986). Intravenous administration of L-NMMA or nitro-arginine methyl ester (L-NAME) results in suppression of NO· release leading to excessive vasconstriction and hypertension (Rees et al. 1989). Even the oral ingestion of L-NAME can induce hypertension (Gardiner et al. 1990). In coronary arteries, only acute pressure increases have been tested to date. These pressure increases result in an immediate loss of endothelium-mediated dilator function and reversal of serotonin-induced dilator response to vasoconstriction. The extent of these changes is dependent on the magnitude and duration of the induced pressure changes (Lamping and Dole 1987). A potentially harmful effect of chronic hypertension on endothelium-mediated control of coronary tone can only be extrapolated from data on peripheral arteries at this time (Linder et al. 1990).

3.5.5 Endothelial Denudation

The extent of endothelium-mediated vascular control is often quantified by comparing responses of endothelium-containing vascular segments to those of endothelium-denuded segments. Denudation can be accomplished by perfusion with enzymes (collagenase), detergents (triton X), or distilled water, or mechanically (e.g., by coronary balloon catheters). These procedures result in the suppression of endothelium-mediated dilator responses, favoring coronary constriction (Nagasawa et al. 1989; Schipke et al. 1985; Tomoike et al. 1989; Yamamoto et al. 1987a, b).

In miniature swine − 3 months after balloon catheter-induced endothelial denudation, in the presence of regional intimal thickening and a regrown endothelial lining − transmembranous Ca^{2+} influx was increased by intracoronary histamine (H_1) stimulation. Histamine elicited focal coronary spasms which were attributed to a regenerated, but dysfunctional endothelium (Yamamoto et al. 1987a, b). Shimokawa et al. (1989b) reported that the regenerated endothelial lining of porcine coronary arteries, after a 3-month healing period, still had a reduced capacity to release EDRF, as indicated by the impaired dilator responses to aggregating platelets and to serotonin. These findings point to a prolonged deficiency of endothelium-dependent coronary dilation even when a regenerated endothelial lining can be demonstrated histologically.

A promotor role of coronary spasm in the progression of coronary atherosclerosis was suggested on the basis of experiments in pigs. A potentiation of coronary constrictor responses and the development of atherosclerosis were observed following a procedure of mechanical endothelial denudation, hypercholesterolemia, and X-ray irradiation (Nagasawa et al. 1989).

When endothelial denudation of peripheral veins by distilled water was carried out in human probands in vivo − thus suppressing the continuous, basal EDRF release and dilator function − an excessive venous constrictor tone appeared (enhanced by platelet activation?) which returned to normal control levels within 15 days (Collier and Vallance 1989).

In balloon catheter-treated patients, platelets in endothelial-denuded segments become immediately activated, adhere to the collagen fibers, aggregate and release platelet compounds such as ADP, serotonin, PAF and TxA$_2$ which elicit focal coronary constriction. Recently, such events during angioplastic widening of coronary and other stenoses were reported which resulted in an increased tendency for local constriction at the stenotic segment until a "neo-intima" was formed (Fischell et al. 1989) and in a somewhat delayed poststenotic small vessel constriction. The latter was apparently related to platelet release products, critically jeopardizing myocardial perfusion even after successful angioplastic interventions (Wilson et al. 1989).

3.6 Endothelin and Coronary Vasomotor Tone

Despite many recent studies on the cardiovascular effects of endothelin (for a review see Yanagisawa and Masaki 1989a, b), only limited information exists on its physiological and pathophysiological significance for coronary control. In particular, little is known about control mechanisms for its abluminal release affecting smooth muscle tone and the significance of luminally released, circulating endothelin, given its extremely low plasma levels in the lower pM range. It remains to be established whether a continuous basal release of endothelin (similar to the continuous EDRF release) contributes significantly to a constrictor resting tone, thus supplementing myogenic or α-adrenergic constrictor coronary tone. Thrombin is one of the few stimulators known to induce endothelin release. This endothelin release is suppressed by EDRF in a cGMP-mediated reaction (Boulanger and Lüscher 1990) though this is controversial. Endothelin may modify adrenergic neurotransmission by pre- and postjunctional mechanisms (Tabuchi et al. 1989; Yanagisawa and Masaki 1989b). Even in the absence of extracellular Ca^{2+}, endothelin increases cytosolic Ca^{2+} and subsequently tension via receptor-dependent, phospholipase C-stimulated phosphatidylinositol disphosphate (PIP$_2$) hydrolysis (Kasuya et al. 1989; Kodama et al. 1989; Marsden et al. 1989). In addition, in the presence of extracellular Ca^{2+} endothelin initiates a transmembrane Ca^{2+} influx (Yanagisawa et al. 1988).

Due to this latter mechanism endothelin-induced constriction can be partially blocked by Ca^{2+} channel blockers.

Infusions of supraphysiological concentrations ($10^{-9} M$) of endothelin cause, apart from myocardial effects, dose-dependent coronary constrictions (Chester et al. 1989; Clarke et al. 1989; Clozel and Clozel 1989; Ezra et al. 1989; Fukuda et al. 1989; Lamping and Eastham 1989; Larkin et al. 1989). The constrictor response is preceded by a short, transient dilator response (Koster et al. 1989), probably attributable to an initial stimulation of PGI_2 and/or EDRF release, as demonstrated in other peripheral beds (de Nucci et al. 1988). The endothelin-induced coronary constrictions reach excessive magnitudes and can induce severe myocardial ischemia resulting in acute heart failure. This potential for severe constriction has been associated in a number of recent publications with the induction of coronary spasm (Franco-Cereceda 1989; Igarashi et al. 1989; Ku 1989; Kurihara et al. 1989; Nakao et al. 1989).

These findings add a further chapter to a large number of reports in the last 30 years suggesting potential causes of coronary spasm and the respective trigger mechanisms, including α-sympathetically induced constriction (Levene and Freeman 1976; Raizner et al. 1980; Tzivoni et al. 1983; Yasue et al. 1976), unbalanced release or action of PGI_2/TxA_2 (Dusting et al. 1979; Vanhoutte and Houston 1985), vasopressin release (Pitt et al. 1982), impaired EDRF release, endothelial dysfunction (Vanhoutte 1987, 1988) and recently, an excessive release of endothelin in patients with vasospastic angina (Toyo-oka et al. 1989). However, any conclusive evidence that endothelin causes spasm is still missing. The induction of severe coronary constriction by exogenously infused endothelin (Clozel and Clozel 1989; Fukuda et al. 1989; Larkin et al. 1989) is certainly different from the action of endogenously released endothelin. This is especially true in the presence of an intact endothelial cell lining which serves as a diffusion barrier for luminally applied endothelin (Pohl and Busse 1989c).

Under ischemic conditions, endothelin concentrations are elevated both in tissue and in circulating blood (Miyauchi et al. 1989; Salminen et al. 1989; Yanagisawa and Masaki 1989b). In infarcted myocardium endothelin concentrations also seem to be elevated. Recently, an increased endothelin receptor expression was found in myocardium under ischemic conditions (Liu et al. 1990). The pathophysiological significance of these findings is not yet clear.

Laminar hemodynamic shear stress not only stimulates EDRF release, but also results in increased messenger RNA for endothelin and stimulation of endothelin release (Yoshizumi et al. 1989). An explanation for these apparently opposing effects is needed. Several pathophysiological conditions – namely cardiogenic shock (Cernacek and Stewart 1989) states associated with endotoxin release (Sugiura et al. 1989), pulmonary hypertension (Cohen and Cunningham 1988), and myocardial infarction (Cohen and Cunningham 1988; Miyauchi et al. 1989; Salminen et al. 1989) — are associated with increased circulating endothelin levels. The specific role of endothelin in coronary vasomotor control under these clinical conditions awaits further elucidation.

4 Humoral Mediators

A number of circulating hormones can modify coronary vasomotor tone using different pathways or second messengers (Fig. 7). Angiotensin II (A II) is washed into the coronary circulation from the systemic circulation, and there is also a local coronary renin-angiotensin system (RAS) (for review see Dzau 1988). There is some evidence that bradykinin plays a role in vascular control. Potentially also involved are circulating norepinephrine and epinephrine, antidiuretic hormone (ADH) = vasopressin (AVP),

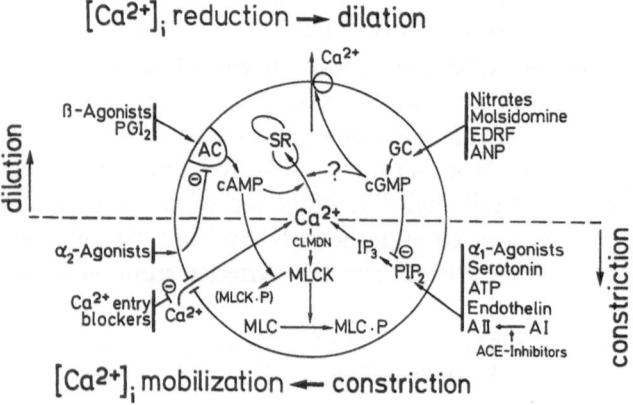

Fig. 7. Main cellular pathways affecting coronary smooth muscle tone. Events leading to relaxation are presented in the *upper half*, and the ones leading to constriction in the *lower half*. Adrenergic β-agonists and PGI$_2$ stimulate adenylate cyclase (*AC*), leading to an increase in cAMP, which in turn results in the reduction of intracellular Ca^{2+} and vascular relaxation. Nitrates, molsidomine, EDRF, and ANP stimulate guanylate cyclase (*GC*), increasing cGMP, thus likewise decreasing Ca^{2+} concentrations and inducing vascular relaxation. Ca^{2+} entry blockers affect voltage-dependent Ca^{2+} channels in vascular smooth muscle and reduce intracellular free Ca^{2+} concentration. In contrast, α$_1$-agonists, angiotensin II, ATP (acting on P$_{2x}$ receptors), endothelin, and serotonin (acting on 5-HT$_2$ receptors) elicit the stimulation of phospholipase C, resulting in an increased diacyl glycerol and inositol trisphosphate (*IP$_3$*) formation. IP$_3$ in turn increases intracellular [Ca^{2+}], thereby initiating vasoconstriction. α$_2$-agonists inhibit AC, decrease cAMP and thus cause constriction. *MLCK*, myosin light chain kinase; *CLMDN*, calmodulin; *PIP$_2$*, phosphatidylinositol diphosphate; *SR*, sarcoplasmic reticulum.

atrial natriuretic peptide (ANP), vasoactive intestinal peptide (VIP), neuropeptide Y (NPY), calcitonin gene-related peptide (CGRP) and, as discussed above, endothelin. Acetylcholine (Ach) is not present in plasma, and thus it is unlikely that it can explain EDRF release at the luminal site (for review see Bassenge and Busse 1988). In autoradiographic binding studies, there was no evidence for the presence of muscarinic receptors on endothelial cells of different vessels (Stephenson et al. 1988). This is surprising because Ach is one of the most powerful stimulators of EDRF production. A transmural diffusion of Ach from the adventitial nerves seems possible (for review see Angus and Cocks 1989), though not very effective in large conduit arteries.

4.1 Peptide Hormones

The evidence showing that peptide hormones play a regulatory role in vascular control is weak. The physiological and pathophysiological significance of CGRP, NPY, and VIP has not been sufficiently analyzed yet. NPY may act as a cotransmitter with local constrictor effects (see below).

Most of these compounds have a dual effect on vascular control. Through their specific receptors on endothelial cells, they stimulate the release of EDRF (sometimes in combination with prostaglandin release) and elicit coronary dilation. On the other hand, they initiate constriction through direct stimulation of their specific receptors on the coronary vasculature (Bassenge and Busse 1988). The subsequent net effect depends on the relative strength of these opposing actions, and under pathophysiological conditions depends in particular on the functional integrity of the endothelial lining.

Surprisingly, in large coronary arteries − in contrast to coronary arterioles − the dilator response of AVP overcomes the constrictor response (Lamping et al. 1989; Myers et al. 1989; Vanhoutte et al. 1984) or may prevent the appearance of a constriction (Monos et al. 1978), though overall coronary resistance is increased by AVP (Heyndrickx et al. 1976; Khayyal et al. 1985). AVP causes vasoconstriction even in the poststenotic, hypoperfused bed (Pantely et al. 1985b) and also in coronary collaterals and collateral-dependent microcirculation (Peters et al. 1989). AVP-induced constrictions in coronary arterioles are potentiated when EDRF effects are inhibited by hemoglobin (Myers et al. 1989). Nevertheless, in physiological concentrations AVP is primarily an antidiuretic hormone preserving the bodies fluid balance ("ADH") rather than acting as a constrictor (Glazier et al. 1989; Nicod et al. 1984; Schmid et al. 1970, 1974) in the coronary bed. Thus, supra-physiological concentrations are usually necessary to elicit significant vasoconstriction and to potentially induce coronary spasm (Pitt et al. 1982), myocardial ischemia (Khayyal et al. 1985), and infarction (Hart and Gokal 1977). ANP can cause endothelium-independent coronary dilation by stimulating particulate guanylate cyclase and a subsequent cGMP increase, an action similar to that of nitrovasodilators (Chu et al. 1987, 1989c). Yet even under pathophysiological conditions (e.g., heart failure) the plasma ANP concentrations necessary for effective vascular control never seem to be reached.

4.2 Circulating Catecholamines

The heart is a major source of circulating catecholamines which spill over from cardiac sympathetic nerves into the coronary circulation. The evidence for β-adrenoceptor-mediated dilator effects on coronary vasomotor tone is weak and originates mainly from three observations: 1) Injections and infusions of norepinephrine or epinephrine usually elicit coronary dilation associated with increases in metabolism (von Restorff and Bassenge 1977). 2) Coronary strip or ring preparations dilate under catecholaminergic stimulation when precontracted (Cohen 1986). 3) β-Adrenergic blockers increase coronary tone and reduce coronary blood flow, though only when associated with a reduction in myocardial metabolism (Vatner and Hintze 1983).

The experimental approaches chosen for such analyses have often been inappropriate. When using infusions, the route of administration is reversed and concentrations are often far beyond physiological levels. Circulating catecholamines are (except for the adrenal release of epinephrine) just the "spill-over fractions" escaping from reuptake mechanisms, and may elicit more or less pronounced endothelium-dependent dilation through stimulation of endothelial α_2-receptors (Cocks and Angus 1983; Matsuda et al. 1985). Local metabolic effects must be experimentally suppressed for such analyses.

Excised vessel preparations are devoid of physiological control mechanisms (nerves, hormones, local factors) and consequently lack resting tone, which has to be replaced experimentally (e.g., by $PGF_{2\alpha}$). In such precontracted, isolated preparations from large conduit arteries, norepinephrine elicits relaxation (Cohen et al. 1984; Cohen 1986) which cannot be observed under in vivo conditions when appropriate controls (especially with regard to simultaneously induced changes in myocardial metabolism) are maintained (for review see Bassenge 1984; Bassenge and Busse 1988).

Adrenergic β-blocker-induced increases in coronary tone are due mainly to reduced myocardial metabolism (i.e., a drop in contractility and heart rate) and only to a small extent to an "unmasking of α-receptor activity " (Vatner and Hintze 1983). Direct neurogenic, β-adrenergic stimulation of coronary dilation (independent of metabolic β_1-mediated effects) is very small (see below).

However, there is convincing evidence that circulating catecholamines exert a regulatory role through α-adrenergic constriction during increased sympathetic stimulation induced by exercise in animals with experimental denervation of defined ventricular regions (Chilian et al. 1986). During submaximal exercise, the α-adrenergic coronary constrictor tone is predominantly due to circulating catecholamines rather than direct neural effects, such that there is no difference in constrictor tone between innervated and denervated regions.

Thus, there is no evidence that humoral catecholamines cause a β-adrenergic coronary dilation in either the large or the small coronary vessels. In contrast, circulating catecholamines exert a significant α-adrenergic constrictor influence on the coronary circulation during exercise.

4.3 Renin-Angiotensin System (RAS)

Angiotensin II (A II) in supraphysiological doses can increase coronary tone both in large coronary conduit vessels and in resistance vessels (Brum et al. 1985; Cohen and Kirk 1973). The formation of A II depends on renin, an enzyme which splits off the decapeptide angiotensin I (A I) from the α_2-globulin angiotensinogen. A I, in turn, is converted into its active form A II (an octapeptide) by an *endothelial* angiotensin converting enzyme (ACE). The carboxypeptidase ACE has a dual function: it activates A I into A II, which induces coronary vasoconstriction, and simultaneously inactivates the nonapeptide bradykinin which attenuates vasoconstriction by an endothelium-mediated dilator mechanism. In addition, A II augments sympathetic effects by a presynaptic facilitation of norepinephrine release (Zimmerman 1978). Inhibitors of ACE can therefore reduce coronary tone by suppressing A II formation and bradykinin inactivation (van Gilst et al. 1987, 1989). Finally, ACE inhibitors can reduce the release of norepinephrine from cardiac sympathetic nerves (Carlsson and Abrahamsson 1989) and, potentially, also its constrictor effects on coronary vasomotor tone.

A number of ACE inhibitors have been shown to reduce coronary tone and increase myocardial perfusion. Under normal conditions the dilator effect of ACE inhibitors is not very pronounced. However, with a low sodium diet or salt depletion (Holtz et al. 1987) or possibly during myocardial ischemia (Ertl 1987), when the RAS is activated and renin secretion stimulated, A II-mediated constrictor tone becomes more pronounced. Administration of ACE inhibitors like enalapril may thus cause coronary dilation, increases in myocardial perfusion and consequently coronary venous oxygen saturation (Holtz et al. 1987).

When coronary perfusion pressure was lowered to induce myocardial hypoperfusion, the activated RAS had a significant constrictor effect, since both a competitive A II antagonist and an inhibitor of ACE attenuated coronary vasomotor tone in this model of myocardial ischemia (Ertl 1987). ACE inhibition by captopril limited infarct size in anesthetized dogs (Ertl et al. 1982) and improved survival in rats subjected to coronary ligation (Pfeffer et al. 1985). However, these positive effects of ACE inhibition by captopril on infarct size were not confirmed in two other studies in conscious dogs (Daniell et al. 1984; Liang et al. 1982). In a recent study, the acute precipitation of poststenotic myocardial ischemia by cardiac sympathetic nerve stimulation in anesthetized dogs (Heusch and Deussen 1983) was not prevented by ACE inhibition with ramiprilat, thus arguing against an anti-ischemic action af ACE inhibitors through the reduction of norepinephrine release (Linder and Heusch 1990).

Finally, two recent clinical studies failed to see any beneficial effect of ACE inhibition in exercise-induced (Gibbs et al. 1989) or pacing-induced (Rousseau et al. 1989) myocardial ischemia in patients with chronic stable angina.

ACE inhibitors can cause dilation of large coronary conduit arteries (Holtz et al. 1987), perhaps by the suppression of a local coronary RAS, i.e., intramural renin-like enzymes and angiotensinogen. Such local systems have been shown in a number of organs (for review see Dzau 1988), but their functional importance, in particular in the coronary circulation, remains to be established.

Thus, there is currently controversial data and no conclusive evidence for a beneficial effect of ACE inhibition in myocardial ischemia, apart from the favorable action during enhanced RAS activation in heart failure and some forms of hypertension (for review see Linder and Heusch 1990).

4.4 Bradykinin

Bradykinin-mediated vasodilation in vascular control is endothelium-mediated by the stimulation of EDRF release and seems to be of minor importance under physiologic conditions. A pentapeptide which substantially potentiates bradykinin-induced coronary dilation did not significantly augment myocardial active (pacing-induced) and reactive hyperemia (Bassenge et al. 1972). The evidence for a potential contribution of bradykinin to coronary vasomotion, mainly under ischemic conditions, is based on the initiation of a significant coronary AV-difference of kininogen, associated with the activation of kallikrein and subsequent release of bradykinin within the coronary bed (Bassenge et al. 1970; Pitt et al. 1969 b). A second indirect line of evidence originates from the effect of recently synthetized, rather specific bradykinin antagonists, which competitively induce vasoconstriction and hypertension (Carbonell et al. 1988; for review see Gavras et al. 1987; Margolius 1989). However, their specific role in the coronary circulation has not yet been tested.

4.5 Prostaglandin-Thromboxane System, Leukotrienes

The eicosanoids are derivatives of arachidonic acid which are formed in platelets and various parts of the vascular wall in a reaction catalyzed by cyclooxygenase (Fig. 8). Prostacyclin (PGI_2) is mainly synthetized in endothelial cells and inhibits platelet aggregation and induces vasodilation. Thromboxane (TxA_2) is mainly synthetized in platelets and only to a very small extent in the vasculature; thromboxane promotes platelet aggregation and vasoconstriction. Cyclooxygenase is the key enzyme for the formation of eicosanoids and can be blocked by substances like indomethacin, acetylsalicyclic acid (aspirin), ibuprofen or diclofenac (for review see Boeynaems 1988).

Since PGI_2 and TxA_2 are very labile, their more stable metabolites 6-keto-$PGF_{1\alpha}$ and TxB_2, respectively − or more recently and appropriately their index metabolites 2,3-dinor-6-keto-$PGF_{1\alpha}$ and 11-dihydro-TxB_2 − are determined in plasma. Platelets, once formed and released from megakaryocytes, are unable to synthetize proteins. Thus when the enzyme cyclooxygenase is blocked, TxA_2 synthesis is discontinued for their life span. However, in the vascular wall de novo synthesis of enzyme proteins

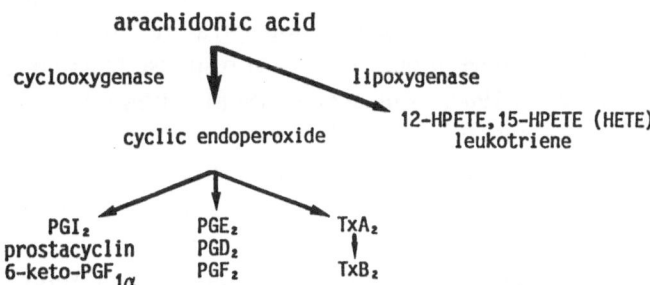

Fig. 8. Biosynthesis of prostacyclin (*PGI_2*) and thromboxane (*TxA_2*) from arachidonic acid catalyzed by cyclooxygenase, and of leukotrienes 12- and 15-HPETE catalyzed by lipoxygenase. Arachidonic acid is cleaved from membrane phospholipids by various phospholipases (mainly A_2) and then transformed into cyclic endoperoxides or 12- and 15-HPETE derivatives (leukotrienes). Cyclic endoperoxides are transformed either into PGI_2 or into PGE_2 and derivatives (PGD_2, PGF_2), or into TxA_2.

renders the inhibition of cyclooxygenase reversible. Thus, the effective balance between the counteracting effects of TxA_2 and PGI_2 is shifted more towards PGI_2. The balance of this system may play a decisive regulatory role in platelet-, peripheral-, and coronary-vasomotor function (Dusting 1984; Moncada et al. 1976).

PGI_2 acts via a receptor-mediated activation of adenylate cyclase. In the vasculature, the subsequent increase in cAMP causes relaxation. In platelets, the increase in cAMP inhibits adhesion, aggregation and the release of pro-aggregatory and vasoconstrictor compounds such as 5-HT, TxA_2, and ADP. The abluminal release of PGI_2 from endothelial cells towards the adjacent vascular smooth muscle cells is small as compared to its luminal release, which causes suppression of platelet activation (Bassenge et al. 1987). A simultaneous EDRF release by a number of stimulator compounds further inhibits platelet activation by augmenting platelet cGMP levels and suppressing the rise of intracellular calcium concentration, which is a prerequisite for platelet activation (Bassenge et al. 1989; Busse et al. 1987). PGI_2 synthesis can be enhanced by ATP, bradykinin, histamine, catecholamines, nifedipine, diltiazem, ACE inhibitors, nitrovasodilators, dipyridamole, and molsidomine (Förster 1980; Schrör 1990; Silberbauer et al. 1982; for review see Boeynaems 1988; Simmet and Peskar 1986). Therefore, it is suggested that a number of vasoactive drugs exert their anti-ischemic action by stimulating PGI_2 formation and release.

Although PGI_2 can induce coronary dilation and TxA_2 coronary constriction, their contribution to the regulation of myocardial perfusion under physiological conditions − when the system is not activated, i. e., in the absence of anesthesia and major surgical trauma − is only minor. Coronary vasomotor tone of functionally intact vessels is not substantially changed by blockade of eicosanoid formation. However, an increase in coronary resistance, a reduction in the diameter of large epicardial conduit arteries, and an increase in venous pressure reflecting a decrease in venous compliance have been reported in response to cyclooxygenase inhibition (Holtz et al. 1984; Münzel et al. 1988). Although the increase in coronary resistance may be specifically related to the use of indomethacin (Edlund et al. 1985), such increases in resistance are also observed in response to other cyclooxygenase blocking agents (Holtz et al. 1984).

The increases in arterial and venous pressure in response to higher concentrations of cyclooxygenase inhibitors augment ventricular afterload and preload. Indomethacin increases myocardial O_2 consumption in concert with an increase in coronary resistance (Friedman et al. 1981). An increase in coronary resistance may contribute to the enhanced frequency of exercise-induced anginal attacks following aspirin in patients suffering from variant angina − even when the rate-pressure product during exercise is lower (Miwa et al. 1981). Apart from increasing pre- and afterload, aspirin may induce a shift from PGI_2 production towards derivatives of the lipoxygenase pathway (i. e., leukotrienes, see below) which do not cause vasodilation (Dusting 1984; Pitt et al. 1983). Conversely, compounds which stimulate PGI_2 synthesis can reduce experimental myocardial infarction and exert additional cytoprotective and antiarrhythmic effects (Jugdutt 1981; Jugdutt et al. 1980; 1981; Schrör et al. 1988; Simmet and Peskar 1986; Thiemermann et al. 1986).

Atherosclerotic vessel segments in vitro exhibit an impaired PGI_2 release upon stimulation (Simmet and Peskar 1986), and higher concentrations of PGI_2 are needed to prevent platelet adhesion at the site of active atherosclerosis (Fitscha et al. 1985). The otherwise surprising increase in biosynthesis of PGI_2 in patients with severe atherosclerosis may compensate for the decreased formation of and response to PGI_2 (FitzGerald et al. 1984).

Thromboxane increases the cytosolic concentration of calcium by means of the formation of inositol 1, 4, 5-trisphosphate (IP_3) and a protein kinase C-mediated mechanism. Thus, in addition to activation and aggregation of platelets, TxA_2 also elicits a decrease in cAMP and contraction of vascular smooth muscle cells (for review see Smith et al. 1980). The constrictor effect of TxA_2 on the coronary vasculature is weaker than originally anticipated from in vitro experiments. The presence of a functionally intact endothelial lining and simultaneous PGI_2 release from the endothelium may partially offset the constrictor effect of TxA_2 (for review see Schrör 1990; Simmet and Peskar 1986). The constrictor effect of platelets in large epicardial coronary arteries may also be mediated by serotonin or other products (for review see Schrör 1990).

It is only in the presence of severe localized coronary stenoses that platelet release products such as TxA_2 and serotonin may be involved in critical coronary flow reductions. Cyclic flow reductions can be observed at the site of severe coronary stenoses, which may be caused both by physical plugging of activated and aggregating platelets and by the constrictor action of their release products, thus initiating a vicious cycle and resulting in critical myocardial blood flow reductions. Such cyclic flow reductions are suppressed by a number of cyclooxygenase inhibitors such as aspirin or indomethacin (Ashton et al. 1986; Folts et al. 1976; Gallagher et al. 1985; Schmitz et al. 1985; Uchida and Murao 1974). In addition to its vasoconstrictor action, TxA_2 is arrhythmogenic (Wainwright and Parratt 1988). As mentioned above, lower concentrations of cyclooxygenase inhibitors preferentially inhibit TxA_2 production. The attenuation of TxA_2-mediated platelet adhesion and aggregation by lower doses of cyclooxygenase inhibitors such as aspirin may contribute to the reduction in the reocclusion rate following coronary bypass surgery, in myocardial reinfarction, and in infarction and death of patients with unstable angina (Hennekens et al. 1988; Lewis et al. 1983; Mustard et al. 1983; Steering Committee of the Physicians' Health Study Research Group 1989)

Elevated levels of TxB_2 can be recovered from the coronary venous effluent of patients with unstable angina during pacing-induced ischemia, whereas 6-keto-$PGF_{1\alpha}$ levels are unaffected (Hirsh et al. 1981). Also, during and immediately following episodes of Prinzmetal angina, TxB_2 levels in coronary sinus plasma are increased (Robertson et al. 1981). Finally, TxB_2 levels in peripheral and coronary sinus plasma of patients with angiographically documented coronary stenoses are significantly elevated as compared to controls, whereas PGI_2 levels are significantly reduced (Kuzuya et al. 1981). Although it is tempting to link an uncontrolled release or action of TxA_2 (or other platelet products like serotonin) in the presence of endothelial damage (Cohen et al. 1983a; Houston et al. 1986; Lam et al. 1987) to the initiation of ischemic episodes, it remains to be established whether the increased TxB_2 levels are cause or consequence of myocardial ischemia. The lack of effectiveness of cyclooxygenase inhibition on frequency and severity of ischemic events in patients with Prinzmetal's angina argues against a significant role for thromboxane, at least in this ischemic syndrome (Chierchia et al. 1982; Robertson et al. 1981).

Thus, the prostaglandin-thromboxane system probably plays no major role in the regulation of coronary blood flow under physiological conditions, although a somewhat increased coronary resistance is observed after inhibition of cyclooxygenase. The prostaglandin-thromboxane system may become more important in pathological states of increased platelet activation as may occur at the site of a severe coronary stenosis.

Therapeutic approaches to mimic or augment the release or action of PGI_2 (vasodilation, inhibition of platelet aggregation, reduction of arrhythmias) have failed so far, probably because severe hypotension is induced before a significant antiplatelet effect can be achieved (for review see Schrör 1990). Using thromboxane synthetase inhibitors, effective suppression of TxA_2 formation (more than 80%) is hard to achieve, and clinical results have been inconclusive. Similarly, high concentrations of TxA_2 receptor antagonists are needed for significant antithrombotic effects (Schumacher et al. 1989), and TxA_2 receptor blockade failed to improve regional myocardial blood flow and function in conscious dogs during exercise-induced ischemia (Thaulow et al. 1989).

Leukotrienes are also derivatives of arachidonic acid, which are synthetized in a lipoxygenase-dependent pathway in granulocytes, macrophages, and mast cells (Fauler and Frölich 1989). Leukotrienes are not involved in the physiological regulation of coronary vasomotor tone. They are probably involved in inflammatory-like processes such as occuring in evolving myocardial infarction or during postischemic reperfusion. Upon exogenous intracoronary infusion, leukotriene C_4 decreases coronary blood flow and regional contractile function in anesthetized dogs; the vasoconstrictor effect persists also in severely stenotic coronary arteries (Nichols et al. 1988; Wargovich et al. 1985). Likewise, intracoronary leukotriene D_4 causes profound coronary vasoconstriction in anesthetized dogs (Fiedler et al. 1984; Panzenbeck and Kaley 1983) and pigs (Fiedler and Abram 1987). The significance of endogenously released leukotrienes is not yet clear and difficult to assess in the complex interplay of multiple mediators, e.g., free radicals, histamine, PAF, and platelet release products. Leukotriene-induced coronary vasoconstriction may contribute to myocardial damage after leukocyte infiltration in evolving myocardial infarction (Evers et al. 1985).

5 Neuronal Mechanisms

Recent studies from different laboratories indicate that the classical concept of maximum coronary dilation during myocardial ischemia is not correct. Ischemic myocardial blood flow is not only determined by the components of extravascular coronary resistance, in particular heart rate and left ventricular pressure (Bretschneider 1967; Heusch and Yoshimoto 1983a; Raff et al. 1972a), but also to a significant extent by coronary constrictor mechanisms. Attenuation of myocardial ischemia with recruitment of coronary dilator reserve is possible (Aversano and Becker 1985; Canty and Klocke 1985; Gorman and Sparks 1982; Heusch and Deussen 1983; Heusch et al. 1987; Pantely et al. 1985a; Seitelberger et al. 1988); however, the mechanisms underlying the presence of vasoconstrictor tone in myocardial ischemia are still poorly understood. Neurogenic mechanisms underlying changes in coronary vasomotor tone have attracted particular interest since they might provide a causal linkage for the acute initiation of myocardial ischemia during exercise and excitement (effort angina: sympathetic mechanism; for review see Heusch 1990) as well as for the apparently paradoxical initiation of myocardial ischemia in resting situations characterized by low myocardial oxygen demand (angina at rest: vagal mechanisms; for review see Heusch and Guth 1989). Both sympathetic and vagal vasomotion may occur at the level of epicardial coronary arteries, in the resistive microcirculation and in the collateral circulation.

5.1 Cholinergic Changes of Coronary Vasomotor Tone

Fibers classified as vagal on the basis of their persistence after sympathectomy and their acetylcholinesterase content have been identified in the coronary arteries of several species including man (Denn and Stone 1976;
Gerova et al. 1979b; Hirsch and Borghard-Erdle 1961). However, the effects of exogenous acetylcholine on the coronary circulation – reaching
the coronary vascular smooth muscle from the luminal (endothelial) rather
than the adventitial site – are still a matter of great controversy. Even
more controversial is the extent and functional significance of the
acetylcholine-mediated vagal regulation of coronary vasomotor tone.

Only part of this controversy can be attributed to the presence or absence of endothelium in various in vitro and in situ vascular preparations.
The pioneering observation by Furchgott and Zawadski (1980) that damage to the endothelial layer reverses acetylcholine-induced relaxation to
constriction in isolated aortic preparations has initiated broad research in
EDRF (covered elsewhere in this issue). Within the framework of this review it should be emphasized that the presence of intact functional endothelium is a prerequisite for epicardial coronary artery dilation in response
to exogenous acetylcholine in anesthetized (Schipke et al. 1985) and conscious dogs (Young et al. 1987) as well as in man (Vita et al. 1990).

Apart from the integrity of endothelium, another confusing problem
arises from the marked negative chronotropic and inotropic effects of
acetylcholine which alter both the metabolic and extravascular components determining coronary blood flow. In experimental studies in which
the heart rate and left ventricular pressure were carefully controlled, cholinergic coronary dilation was consistently observed in response to exogenous acetylcholine (Feigl 1969; Schipke et al. 1985), electrical vagal stimulation (Feigl 1969), and reflex vagal activation through baroreceptors
(Hackett et al. 1972; Ito and Feigl 1985a), chemoreceptors (Hackett et al.
1972; Ito and Feigl 1985b), and ventricular receptors (Zucker et al. 1987).
However, all of these studies were performed in dogs, and in each instance
the observed coronary dilation was only transient. In these canine preparations, the observed increases in coronary blood flow were most probably
based on a dilation of the resistive coronary microcirculation since epicardial coronary resistance is only small in the absence of a proximal stenosis
(Fam and McGregor 1968; Kelley and Feigl 1978). There is also a dilation
of canine epicardial coronary arteries in response to exogenous acetylcholine (Schipke et al. 1985; Young et al. 1987) which is slower in onset and
more sustained than the small vessel dilation (Cox et al. 1983). However,
during electrical stimulation of both cervical vagi there is no dilation of
canine epicardial coronary arteries (Gerova et al. 1979b). Intracoronary

acetylcholine increases blood flow more in the subendocardium than in the subepicardium, whereas electrical vagal stimulation induces uniform dilation across the left ventricular wall in dogs (Reid et al. 1985). In summary, all of these more recent and carefully controlled studies in canine preparations demonstrate unequivocally cholinergic coronary dilation and provide absolutely no evidence that a cholinergic coronary constrictor mechanism may be involved in the initiation of myocardial ischemia.

However, there are obvious species differences. In conscious calves, there is a biphasic response with a transient coronary constriction followed by a more sustained dilation of epicardial and small resistive coronary arteries in response to intracoronary acetylcholine (Young et al. 1987, 1988a). In chronically instrumented, sedated baboons, the predominant response to intracoronary acetylcholine is intense coronary constriction which is unaffected by α- or β-blockers but abolished by atropine (Young et al. 1987). However, it was recently pointed out that in anesthetized baboons coronary constriction is secondary to acetylcholine-induced decreases in myocardial oxygen consumption at high doses whereas the predominant response to lower doses of acetylcholine is coronary dilation (van Winkle and Feigl 1989).

In isolated human vascular preparations, cholinergic coronary constriction is observed regardless of the integrity of the endothelium (Kalsner 1985). The few existing studies on the effects of acetylcholine in the human coronary circulation in situ are inconsistent. In angiographically normal coronary arteries, the predominant response to intracoronary acetylcholine appears to be dilation although constriction is also observed (Ludmer et al. 1986). In atherosclerotic segments, however, the predominant response appears to be constriction (Drexler et al. 1989; Fish et al. 1988; Hodgson and Marshall 1989; Horio et al. 1986; Ludmer et al. 1986; Vita et al. 1990; Werns et al. 1989) (Fig. 9). It has to be kept in mind, though, that these studies do not demonstrate cholinergic regulation of coronary vasomotor tone, but responses to exogenous acetylcholine reaching the coronary vascular smooth muscle from the luminal (endothelial) rather than the adventitial site. Thus, at present there is neither evidence for a physiological role of vagal regulation of coronary blood flow nor for a pathophysiological role in myocardial ischemia in man.

5.2 Adrenergic Changes of Coronary Vasomotor Tone

The morphology of cardiac sympathetic nerves in several species including man has been well described (Borchard 1978; Denn and Stone 1976; Holmgren et al. 1985), and their reflex activation during exercise and ex-

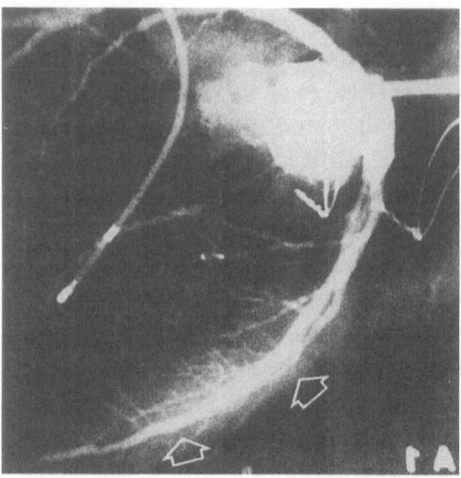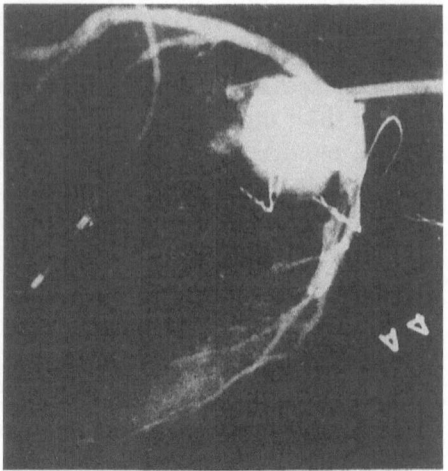

Fig. 9. Angiographic demonstration of the marked constriction of a coronary artery with diffuse atherosclerosis in response to intracoronary infusion of acetylcholine. From Fish et al. (1988) by permission of Rockefeller University Press.

citement is beyond doubt (for review see Feigl 1983; Heusch et al. 1986). The activation of cardiac β-adrenoceptors mediates an increase in heart rate and myocardial inotropism. The resulting increase in myocardial oxygen demand is adequately matched by an augmented oxygen supply after metabolic dilation of the coronary vasculature under normal conditions (Berne et al. 1965; Miller et al. 1979). However, the direct effect of the sympathetic neurotransmitter norepinephrine on the coronary vascular smooth muscles is vasoconstriction which is mediated by activation of α-adrenoceptors (Berne 1958). Even under normal conditions in the presence of a substantial coronary dilator reserve, α-adrenergic constriction acts to limit metabolic coronary dilation by about 30%. Thus myocardial oxygen extraction increases together with coronary blood flow during sympathetic activation in order to match oxygen supply to the increased myocardial oxygen demand (Feigl 1975; Mohrman and Feigl 1978; Murray and Vatner 1979).

Some studies in isolated, saline-perfused rat (Richardt et al. 1987, 1989) and guinea pig hearts (Schrader et al. 1977) suggest that local metabolic factors such as adenosine exert a modulating, inhibitory influence on adrenergic neurotransmission both at presynaptic (Richardt et al. 1987) and postsynaptic sites (Schrader et al. 1977). However, evidence for such an inhibitory effect of adenosine on adrenoceptor mediated increases in myocardial inotropic state (Seitelberger et al. 1984, Schipke et al. 1987) and coronary constrictor tone (Heusch et al. 1986) could not be obtained under more physiological circumstances in canine hearts in situ.

Whether or not α-adrenergic coronary constriction is powerful enough to limit coronary blood flow under ischemic conditions, i.e., in the presence of exhausted coronary dilator reserve, is a matter of great controversy (for review see Feigl 1987; Heusch 1990). Also controversial is whether α-adrenergic coronary constriction, when still effective during myocardial ischemia, exerts a beneficial or deleterious influence on the ischemic myocardium. These controversies arise from the use of different animal preparations, from clinical observations in different types of angina, drugs and procedures used for sympathetic activation, and finally to a large part from the use of an ever increasing number of α-adrenoceptor agonists and antagonists with significantly different pharmacological properties.

5.2.1 β-Adrenergic Coronary Dilation

A direct β-adrenergic coronary dilation, independent of metabolic factors, has been demonstrated both in large epicardial and small resistive coronary arteries in situ (Hamilton and Feigl 1976; Klocke et al. 1965; McRaven et al. 1971; Vatner et al. 1982, 1986; von Restorff and Bassenge 1977). The direct coronary dilation of epicardial coronary arteries is mediated by both β_1- and β_2-receptors in dogs (Nakane et al. 1988; Vatner et al. 1982) and calves (Vatner et al. 1986). In canine large coronary arteries in vitro the adrenergic vasodilation is predominantly mediated by vascular β_1-adrenoceptors (Nakane et al. 1988). In the small resistive coronary arteries of dogs β_2-adrenoceptors predominate (Murphree and Saffitz 1988). The physiological significance of β-adrenoceptor mediated coronary dilation during sympathetic activation, however, appears to be minimal (Hamilton and Feigl 1976; McRaven et al. 1971). In epicardial coronary arteries, β-adrenoceptor activation may also exert indirect dilatory effects through an ascending mechanism (Holtz et al. 1983) secondary to increases in coronary blood flow after metabolically mediated small vessel dilation. Lack of this indirect dilatory effect may be responsible for the large vessel constriction observed after pharmacological β-blockade, rather than unopposed α-adrenergic constrictor tone (Vatner and Hintze 1983). Mature canine coronary collateral vessels express a mixed population of β_1- and β_2-adrenoceptors mediating relaxation in vitro (Feldman et al. 1989) and in vivo (Maruoka et al. 1987).

5.2.2 α-Adrenergic Coronary Vasomotion

5.2.2.1 α-Adrenoceptor Subtypes: Classification, Agonists, and Antagonists

α-Adrenoceptors are classified with respect to their location as presynaptic and postsynaptic and with respect to their pharmacological properties as α_1- and α_2-adrenoceptors (Hoffman and Lefkowitz 1980; Homcy and Graham 1985; Langer 1981; McGrath 1982; Starke 1981; van Zwieten and Timmermans 1983). Traditionally, postsynaptic and α_1-adrenoceptors have been regarded as identical, mediating the constriction of all kinds of vascular beds, including the coronary circulation. Conversely, presynaptic and α_2-adrenoceptors have been regarded as identical, mediating a feedback inhibition of neuronal norepinephrine release from sympathetic nerve terminals. However, this classification is too simplistic since recent studies indicate the presence of both α_1- and α_2-adrenoceptors at both presynaptic sympathetic nerve terminals and postsynaptic sites of different cell types.

Norepinephrine and epinephrine act on both α_1- and α_2-adrenoceptors. Methoxamine and phenylephrine are selective agonists for α_1-adrenoceptors (Kobinger and Pichler 1981; van Meel et al. 1981). However, phenylephrine simultaneously activates β-adrenoceptors and can therefore be used as a selective α_1-adrenoceptor agonist only in the presence of β-blockade (van Meel et al. 1981). BHT 920, BHT 933 (azepexole), and UK-14,304 are selective agonists for α_2-adrenoceptors (Kobinger and Pichler 1981; van Meel et al. 1981). Clonidine, which is occasionally used as an α_2-adrenoceptor agonist, has a lesser selectivity for α_2-adrenoceptors (Kobinger and Pichler 1981) and is only a partial agonist at presynaptic α_2-adrenoceptors (van Meel et al. 1981). Phentolamine is a nonselective α-adrenoceptor antagonist. Prazosin is a highly selective α_1-adrenoceptor antagonist (Cambridge et al. 1977). Yohimbine, rauwolscine, and particularly idazoxan are selective α_2-adrenoceptor antagonists (Doxey et al. 1983; Weitzell et al. 1979). The postsynaptic α_2-selectivity of SKF 104,078 proposed by Ruffolo et al. (1987) was not confirmed in other studies (Connaugthon and Docherty 1988, 1990). The use of phenoxybenzamine in studies on α-adrenergic mechanisms is problematic since it induces complete and irreversible blockade of α_1-adrenoceptors, but only incomplete blockade of α_2-adrenoceptors (Constantine and Lebel 1980).

5.2.2.2 Presynaptic α-Adrenoceptors

The presence and functional role of presynaptic α-adrenoceptors on cardiac sympathetic nerve terminals is well established. However, the augmentation of neuronal norepinephrine release (Heyndrickx et al. 1984; Langer et al. 1977) and of myocardial (Dai et al. 1986; Heyndrickx et al. 1984; Langer et al. 1977), peripheral vascular (Saeed et al. 1982), and coronary vascular (Cohen et al. 1983b) responses to sympathetic activation after nonselective α-adrenoceptor blockade by phentolamine does not distinguish between presynaptic α_1- and α_2-adrenoceptors. Selective activation of presynaptic α_2-adrenoceptors reduces neuronal norepinephrine release (Yamaguchi et al. 1977) and the resulting tachycardia (van Meel et al. 1981) during cardiac sympathetic nerve stimulation. Their blockade enhances neuronal norepinephrine release during cardiac sympathetic nerve stimulation (Dart et al. 1984) and exercise (Heyndrickx et al. 1984) as well as the resulting increases in heart rate and contractility (Guth et al. 1990; Heyndrickx et al. 1984). Some studies also suggest the presence of α_1-adrenoceptors on cardiac sympathetic nerve terminals of several species, mediating a feedback inhibition of neuronal norepinephrine release during cardiac sympathetic nerve stimulation, similar to that mediated by α_2-adrenoceptors (Cavero et al. 1979; Constantine et al. 1978; Kobinger and Pichler 1980). However, the presence and functional importance of presynaptic α_1-adrenoceptors for the hemodynamic responses to exercise in conscious dogs is controversial. Heyndrickx et al. (1984) demonstrated that the arteriovenous norepinephrine gradient across the myocardium and the increases in heart rate and left ventricular dP/dt during exercise were potentiated by intravenous and intracoronary administration of the selective α_2-antagonist yohimbine, whereas the intravenous administration of the selective α_1-antagonist prazosin caused no such potentiation. In contrast, in a recent study by Guth et al. (1990) the exercise-induced increases in heart rate and regional and global left ventricular contractility were augmented not only by the selective α_2-antagonist idazoxan, but also by the selective α_1-antagonist prazosin, when the prazosin-induced hypotension was prevented. The potentiation of exercise-induced increases in heart rate and contractility by idazoxan and prazosin was eliminated by β-blockade, suggesting a functionally important role for both α_1- and α_2-adrenoceptors in the control of neuronal norepinephrine release during the sympathetic activation induced by exercise.

5.2.2.3 α-Adrenoceptors on Endothelium and Platelets

Endothelial cells of large canine and porcine coronary arteries carry α_2-adrenoceptors. Their activation by norepinephrine causes release of an endothelium derived relaxant factor which in turn attenuates norepinephrine-induced vasoconstriction mediated by vascular α_1-adrenoceptors (Cocks and Angus 1983). The activation of endothelial α_2-adrenoceptors contributes to the polarity of the vascular wall in canine femoral and rat tail arteries in that constrictions induced by the luminal application of BHT 933 or UK-14,304 are less pronounced than those by adventitial application (Matsuda et al. 1985). In rabbit carotid arteries the endothelium attenuates the vasoconstriction induced by sympathetic nerve activation (Tesfamariam and Cohen 1988), metabolizes norepinephrine, physically limits norepinephrine overflow into the vessel lumen, and finally inhibits the release of norepinephrine from sympathetic nerve terminals (Cohen and Weisbrod 1988). Shear stress at the endothelial surface of isolated rabbit carotid arteries activates the endothelial cells and causes them to attenuate adrenergic vasoconstriction (Tesfamariam and Cohen 1988). A shear stress-dependent, endothelium-mediated dilation has also been demonstrated in epicardial coronary arteries of conscious dogs at increased coronary blood flow (Holtz et al. 1984) and may act to modulate coronary vasomotion when metabolic dilation and α-adrenergic constriction compete during sympathetic activation. Conversely, loss of endothelial function in atherosclerotic coronary arteries may predispose to enhanced α-adrenergic coronary vasoconstriction (Heistad et al. 1984; Nabel et al. 1988a; Rosendorff et al. 1981).

Human platelets carry α-adrenoceptors which in radioligand binding studies have shown to be α_2-adrenoceptors (Motulsky et al. 1983, 1984). Quinidine (Motulsky et al. 1984) and some but not all Ca^{2+} antagonists (Motulsky et al. 1983) compete for these binding sites. The activation of α_2-adrenoceptors mediates the aggregation of human platelets in response to norepinephrine and epinephrine. However, human platelets also appear to carry α_1-adrenoceptors at which phenylephrine acts as a partial agonist (Grant and Scrutton 1979). Platelets may play a significant role both in the development of coronary atherosclerosis and in the precipitation of myocardial ischemia (Fuster et al. 1985). The activation of platelets by catecholamines may be a contributing factor for enhanced platelet aggregation in patients with coronary artery disease during sympathetic activation such as exercise (Kumpuris et al. 1980).

5.2.2.4 Coronary Vascular α-Adrenoceptors

Compared to cutaneous and skeletal muscle vasculature, there are only minor α-adrenergic constrictor effects in canine coronary vessels during norepinephrine infusion and sympathetic nerve stimulation (Mark et al. 1972). During supramaximal cardiac sympathetic nerve stimulation in the presence of β-blockade the increase in coronary resistance amounts to only 20% – 30% in anesthetized dogs (Heusch et al. 1984; Kelley and Feigl 1978). Both postsynaptic α_1- and α_2-adrenoceptors are present on coronary vascular smooth muscle cells of various species. However, data on the quantitative contribution of vascular α_1- and α_2-adrenoceptors to the constriction of large epicardial, small resistive and coronary collateral vessels are still somewhat controversial.

5.2.2.4.1 Epicardial Coronary Arteries. Epicardial coronary resistance contributes only about 5% of the total coronary resistance in the absence of coronary stenosis (Fam and McGregor 1968; Kelley and Feigl 1978). Furthermore, the increase in epicardial coronary resistance during cardiac sympathetic nerve stimulation in the presence of β-blockade is even less pronounced than that of total coronary resistance (Heusch et al. 1984; Kelley and Feigl 1978). Conversely, the decrease in epicardial coronary diameter during cardiac sympathetic nerve stimulation in the presence of β-blockade is less than 5% of the resting diameter (Gerova et al. 1979a; Heusch et al. 1984).

In isolated epicardial canine coronary arteries, adrenergic vasoconstriction appears to be mediated exclusively by α_1-adrenoceptors (Nakane and Chiba 1987; Rimele et al. 1983; Toda 1986). However, in isolated epicardial human and monkey coronary arteries, both α_1- and α_2-adrenoceptors are involved in norepinephrine-induced constriction (Toda 1986). In epicardial coronary arteries of anesthetized β-blocked dogs in situ, there is a vasoconstrictor response to the intracoronary infusion of the selective α_1-agonist methoxamine and a prazosin-sensitive constriction during cardiac sympathetic nerve stimulation, whereas intracoronary infusion of the selective α_2-agonist BHT 920 induces no epicardial coronary constriction (Heusch et al. 1984). In conscious calves with β-blockade, however, equivalent reductions in epicardial coronary diameter by about 5% are induced by intracoronary infusion of the selective α_1-agonist phenylephrine and the selective α_2-agonist BHT 920, which are abolished by the selective α_1-antagonist prazosin or the selective α_2-antagonist rauwolscine, respectively (Young et al. 1988b). Ca^{2+} antagonists inhibit the α-adrenoceptor mediated constriction of epicardial coronary arteries in vitro (Rimele et al. 1983; Toda 1986) and in situ (Heusch et al. 1984).

5.2.2.4.2 Coronary Resistive Vessels. Intravital microscopic evaluation of the coronary microcirculation in cats demonstrated a nonuniform constrictor response to α-adrenoceptor activation with increasing doses of exogenous norepinephrine and frequencies of cardiac sympathetic nerve stimulation. A significant constriction of arterial and larger arteriolar segments with a resting diameter of more than 100 μm, and a dilation of arterioles with a resting diameter of less than 100 μm, occurred during α-adrenoceptor activation in the presence of β-blockade (Chilian et al. 1989). These recent findings indicate a different coronary vascular site for metabolic dilation and α-adrenergic constriction and a redistribution of coronary vascular resistance towards larger coronary vessels during α-adrenoceptor activation. The simultaneous constriction and dilation of various coronary vascular segments may explain the weak net constrictor effect of α-adreno-ceptor activation in the coronary circulation (see above) as compared with skeletal muscle in which α-adrenergic vasoconstriction occurs in a broad range of resistance vessels of different caliber (Faber 1988).

Holtz et al. (1982) were the first to demonstrate that, in anesthetized dogs, the norepinephrine-induced increase in coronary resistance was sig-nificantly more attenuated by the selective α_2-antagonist rauwolscine than by the selective α_1-antagonist prazosin. The predominance of α_2-adreno-ceptors in mediating coronary vasoconstriction in response to intracoro-nary α-adrenoceptor agonists (Chen et al. 1988; Heusch et al. 1984) and cardiac sympathetic nerve stimulation (Heusch et al. 1984; Saeed et al. 1985), was later confirmed in anesthetized canine preparations. However, in conscious dogs the increase in end-diastolic coronary resistance in re-sponse to intravenous norepinephrine appeared to be mediated by both α_1- and α_2-adrenoceptors since it was attenuated by the α_1-antagonist prazosin and the α_2-antagonist rauwolscine to almost the same extent (Woodman and Vatner 1987). Whereas intravenous administration of nor-epinephrine increases coronary perfusion pressure and ventricular after-load and causes reflex sympathetic withdrawal, regional intracoronary in-fusion of norepinephrine in conscious, chronically instrumented dogs re-vealed once more a strong predominance of α_2-adrenoceptors mediating the increase in end-diastolic coronary resistance, which was abolished by the selective α_2-antagonists rauwolscine and idazoxan (Seitelberger et al. 1986, 1988). α-Adrenoceptor mediated coronary constriction in anesthe-tized dogs can functionally be antagonized by the Ca^{2+} antagonists nifedipine (Heusch et al. 1984) and felodipine (Ehring and Heusch 1990).

The presence of α_1- and α_2-adrenoceptors has also been shown in the coronary circulation of isolated guinea-pig hearts. However, their distribu-tion and functional importance in mediating constrictor responses to neu-ronal and humoral norepinephrine has not been evaluated (Decker and Schwartz 1985). In pigs anesthetized with chloralose, enflurane or iso-flurane, α-adrenoceptor mediated coronary constriction is even sparser than in dogs, with virtually no responses to intracoronary infusion of the α_1-agonist methoxamine and only weak and variable coronary constrictor responses to the α_2-agonist BHT 933 (Schulz et al. 1990). Both α_1- and α_2-adrenoceptor mediated constrictions have been demonstrated in hu-man forearm vessels in situ (Jie et al. 1984, 1987). However, the presence, distribution, and functional importance of α_1- and α_2-adrenoceptors in the human coronary circulation remain to be established.

5.2.2.4.3 Collaterals. The vasomotor response of coronary collateral vessels to α-adrenoceptor activation has so far only been studied in dogs. Collateral vasoconstriction in response to the α_1-agonist phenylephrine was suggested by an acute study in fibrillating dog hearts during cardiopulmonary bypass (Sink et al. 1979). However, in this study phenylephrine was infused systemically and the effects of phenylephrine on coronary perfusion pressure, the extravascular component of coronary resistance, and coronary vasomotor tone in the nonischemic and ischemic myocardium could not be adequately separated from its effects on the collateral circulation. In another study (Maruoka et al. 1987), no α-adrenoceptor mediated constriction of mature collaterals or native coronary vessels was reported in response to intracoronary norepinephrine. However, a marked constriction in response to intracoronary infusion of the selective α_2-agonist BHT 920 was observed. It remains unclear why, in this study, BHT 920 induced a constrictor response that was not observed in response to norepinephrine, which acts on coronary vascular α_1- and α_2-adrenoceptors.

The absence of any constrictor response to α_1- and α_2-adrenoceptor activation was reported from carefully controlled experiments on the native immature (Hautamaa et al. 1987) and mature (Harrison et al. 1986; Hautamaa et al. 1989) canine collateral cirulation in situ. The lack of responsiveness to α-adrenoceptor activation was confirmed after isolation of the collateral vessels in vitro (Harrison et al. 1986).

Because of the apparent lack of α-adrenergic vasoconstriction in the collateral vessels themselves, collateral blood flow during sympathetic activation is determined by the driving pressure gradient across the collaterals. This driving pressure gradient is reduced during sympathetic activation by a simultaneous metabolic vasodilation in the terminal vascular bed of the nonischemic donor side and an α-adrenoceptor mediated vasoconstriction in the terminal vascular bed of the ischemic recipient side (Busch et al. 1988).

5.2.2.4.4 Regional and Transmural Distribution of α-Adrenergic Coronary Constriction. During electrical stimulation of different cardiac sympathetic nerves in anesthetized, β-blocked dogs there are marked regional variations in the coronary constrictor response in different regions of the left and right ventricle (Haws et al. 1987; Rinkema et al. 1982). Electrical stimulation of the left ventrolateral cardiac sympathetic nerve induces a more pronounced vasoconstrictor response in the subepicardial layers of the posterolateral myocardium than in the subendocardial layers (Haws et al. 1987; Johannsen et al. 1982), resulting in an improved ratio of subendocardial to subepicardial blood flow. However, it should be emphasized that this improved ratio was still associated with some decrease in subendocardial blood flow (Johannsen et al. 1982). The coronary constrictor response to intracoronary α_1-adrenoceptor activation with phenylephrine and α_2-adrenoceptor activation with BHT 933 is transmurally homogeneous, with no redistribution of blood flow occurring in nonischemic myocardium (Chen et al. 1988).

5.2.2.4.5 α-Adrenergic Coronary Constrictor Tone at Rest. Significant α-adrenergic coronary constrictor tone at rest was suggested by several studies in anesthetized (Schwartz and Stone 1977) and conscious dogs (Holtz et al. 1977b; Vlahakes et al. 1982). A further reduction in coronary vasomotor tone resulted from nonselective α-adrenoceptor blockade by phentolamine in conscious dogs with β-blockade and maximal coronary dilation by adenosine (Vlahakes et al. 1982). Reduction of α-adrenergic coronary constrictor tone at rest resulting in a preferential improvement in subendocardial blood flow was reported in anesthetized, vagotomized dogs after surgical left stellectomy (Schwartz and Stone 1977) and in conscious dogs after acute chemical regional myocardial sympathectomy using 6-hydroxydopamine (Holtz et al. 1977b). Obviously it is critical for all these studies to confirm true resting basal conditions. Such a true resting state certainly does no exist in chloralose-anesthetized dogs (Schwartz and Stone 1977), and heart rate or maximum dP/dt were relatively high in the two studies with conscious dogs (Holtz et al. 1977b; Vlahakes et al. 1982). Furthermore, nonspecific coronary vasodilation secondary to toxic damage by 6-hydroxydopamine cannot be excluded (Holtz et al. 1977b). In contrast to the previously mentioned studies, Chilian et al. (1981), in well controlled experiments, were unable to demonstrate any α-adrenergic coronary constrictor tone in conscious dogs under true resting conditions in a myocardial region sympathectomized 2 weeks before by perivascular incisions and topical application of phenol. These discrepant experimental results are difficult to reconcile at present, in particular since all interventions eliminating one control mechanism, i.e., resting sympathetic constrictor tone, may induce a compensatory increase in the functional significance of other control mechanisms, e.g., myogenic tone, and such a compensatory mechanism may become more prominent over time. In man, there appears to be some evidence for resting α-adrenergic coronary constrictor tone, since normally innervated patients are characterized by a higher resting coronary resistance and a higher coronary arteriovenous oxygen difference than cardiac transplanted patients; this difference is abolished by nonselective α-blockade with phentolamine (Orlick et al. 1978). However, the conclusions of this study are largely based on the use of coronary sinus thermodilution, which may not be suitable for detecting small changes in coronary blood flow. Indeed, a recent study using an intracoronary Doppler catheter for flow velocity measurements and quantitative coronary angiography for diameter measurements revealed only negligible α-adrenoceptor mediated coronary resting tone in both denervated transplant patients and normally innervated patients (Hodgson et al. 1989).

5.2.2.4.6 Reflex Regulation of α-Adrenergic Coronary Constriction. An α-adrenoceptor mediated increase in coronary resistance is part of the cardiac response to carotid baroreceptor reflex activation (Feigl 1968; Ito and Feigl 1985a). Carotid chemoreceptor activation induces a biphasic coronary vasomotor response with early dilation and late constriction. In conscious dogs with intracarotid injection of nicotine, the dilator response has been suggested to involve the withdrawal of α-adrenergic coronary constrictor tone secondary to an increased depth of inspiration (Vatner and McRitchie 1975). However, an equivalent attenuation of this coronary dilator response by α-adrenoceptor blockade in normally innervated and cardiac-denervated conscious dogs excludes a significant role of cardiac sympathetic nerves in the coronary dilation during carotid chemoreceptor activation (Nagata et al. 1988). In contrast to dogs, the pulmonary inflation reflex has almost no physiological significance in the coronary circulation of conscious man (Wilson et al. 1988). With controlled ventilation, a significant α-adrenoceptor mediated coronary constrictor response is apparent in conscious dogs during carotid chemoreceptor activation (Murray et al. 1984).

Several central nervous sites from which α-adrenergic coronary constriction can be elicited have been identified in anesthetized, β-blocked cats (Bonham et al. 1987a, b). However, a physiological or pathological role for these central nervous sites in regulating coronary blood flow or inducing myocardial ischemia remains to be established.

5.2.2.4.7 α-Adrenergic Coronary Constriction During Exercise. In the absence of β-blockade, nonselective α-blockade by phentolamine (Heyndrickx et al. 1984), as well as selective α_1-blockade by prazosin (Gwirtz et al. 1986; Strader et al. 1988) and selective α_2-blockade by yohimbine (Heyndrickx et al. 1984), augment the decrease in coronary vascular resistance during exercise. However, as discussed previously, in the absence of β-blockade this effect of systemic (Guth et al. 1990; Heyndrickx et al. 1982, 1984) or regional intracoronary (Gwirtz et al. 1986; Heyndrickx et al. 1984; Strader et al. 1988) α-blockade can be caused by presynaptic disinhibition of neuronal norepinephrine release resulting in enhanced metabolic coronary vasodilation as well as by the attentuation of α-adrenergic coronary vasoconstriction. Furthermore, systemic phentolamine and prazosin induce hypotension resulting in an additional baroreflex-mediated sympathetic activation (Heyndrickx et al. 1984). The quantitative contribution of baroreflex-mediated sympathetic activation, presynaptic disinhibition of

norepinephrine release and attenuation of α-adrenergic coronary vasoconstriction to the potentiation of exercise-induced coronary vasodilation
by α-blockade is difficult to assess. Nevertheless, studies in conscious dogs
with β-blockade demonstrate that α-adrenergic coronary vasoconstriction
limits exercise-induced metabolic coronary dilation (Gwirtz et al. 1986;
Murray and Vatner 1979). Gwirtz et al. (1986) even reported a limitation
of exercise-induced increases in cardiac function by α-adrenergic coronary
vasoconstriction. However, no augmentation of cardiac performance occurred in two other studies using treadmill exercise in β-blocked, conscious
dogs after systemic phentolamine (Murray and Vatner 1979) or selective
α_1- and α_2-blockade (Guth et al. 1990). This apparent difference could be
due to different exercise workloads resulting in a different recruitment of
coronary arteriovenous oxygen difference in matching oxygen supply to
myocardial oxygen demand. The contribution of α_1- and α_2-adrenoceptors
to coronary vasoconstriction during exercise has not been assessed in a β-
blocked preparation so far.

α-Adrenergic coronary constrictor tone during submaximal exercise is
exerted predominantly by circulating catecholamines and not by local catecholamine release from cardiac sympathetic nerves, since there was no difference in myocardial blood flow between an innervated and a sympathectomized region in exercising dogs during β- and combined α- and β-block-
ade (Chilian et al. 1986). There was also no difference in the transmural
distribution of myocardial blood flow between the innervated and the sympathectomized region, suggesting transmurally homogeneous α-adrenergic
coronary vasoconstriction (Chilian et al. 1986).

In contrast, Huang and Feigl (1988) described a transmurally nonuniform α-adrenergic coronary vasoconstriction which acted to maintain a
homogenous transmural blood flow distribution during exercise by preferential α-adrenergic vasoconstriction in the subepicardium. However, apart
from the problematic use of phenoxybenzamine in this study, only small
differences in normalized and interpolated transmural blood flow ratios
between an intact and a phenoxybenzamine-treated region were demonstrated. Since total blood flow was still higher in the phenoxybenzamine-
treated region, a beneficial role of α-adrenergic coronary vasoconstriction
for transmural myocardial perfusion during exercise seems not to be well
substantiated.

5.3 Sympathetic Mechanisms in Myocardial Ischemia

5.3.1 β-Adrenergic Mechanisms in Myocardial Ischemia

The β-adrenergic mechanisms contributing to myocardial ischemia appear
to be essentially indirect – through an unfavorable redistribution of coro-

nary blood flow away from the ischemic subendocardium, i. e., through a collateral and a transmural steal mechanism. In conscious dogs with a severe single vessel stenosis, produced by the progressive narrowing of an ameroid constrictor on the left circumflex coronary artery, resting blood flow and function of the poststenotic myocardium are not compromised by virtue of the development of an extensive collateral circulation. During the enhanced sympathetic activity of treadmill exercise, however, ischemia is induced in the poststenotic myocardium. Subendocardial blood flow of the ischemic region is drastically reduced while blood flow to the overlying subepicardium is increased as is blood flow to all transmural layers of the nonischemic myocardium. Regional myocardial systolic wall thickening is severely reduced in the ischemic region while it increases in the nonischemic region. β-Adrenoceptor blockade with atenolol attenuates the exercise-induced increases in heart rate, left ventricular dP/dt, and regional function of the nonischemic region. Also, increases in blood flow in the nonischemic control zone and in the poststenotic subepicardium are reduced. However, subendocardial blood flow in the poststenotic myocardium is increased, resulting in increased regional systolic wall thickening (Matsuzaki et al. 1984). The beneficial effects of atenolol in exercise-induced myocardial ischemia are due exclusively to the attenuation of the increase in heart rate. When the reduction in heart rate by atenolol is prevented by atrial pacing, ischemic regional myocardial blood flow and function are even slightly reduced compared to the untreated situation, possibly due to an unmasking of α-adrenergic constriction in the ischemic coronary microcirculation (Guth et al. 1987a). Conversely, attenuation of exercise-induced tachycardia by a selective bradycardic agent substantially improves blood flow and function in the ischemic myocardium (Guth et al. 1987b). The hemodynamic severity of a dynamic coronary stenosis can be reduced by β-blockade. Atenolol, metoprolol and propranolol induce an autoregulatory decrease in flow to nonischemic regions which increases poststenotic coronary perfusion pressure, thereby effectively reducing stenotic resistance and improving blood flow to ischemic regions (Buck et al. 1981).

5.3.2 α-Adrenergic Coronary Constriction
in Experimental Myocardial Ischemia

Bassenge et al. (1967) first observed a shift from metabolic coronary dilation to coronary constriction during intracoronary norepinephrine infusion, when coronary perfusion pressure in anesthetized dogs was progressively reduced (Fig. 10a). The dependence of norepinephrine-induced coronary vasomotion on preexisting coronary vasomotor tone was not

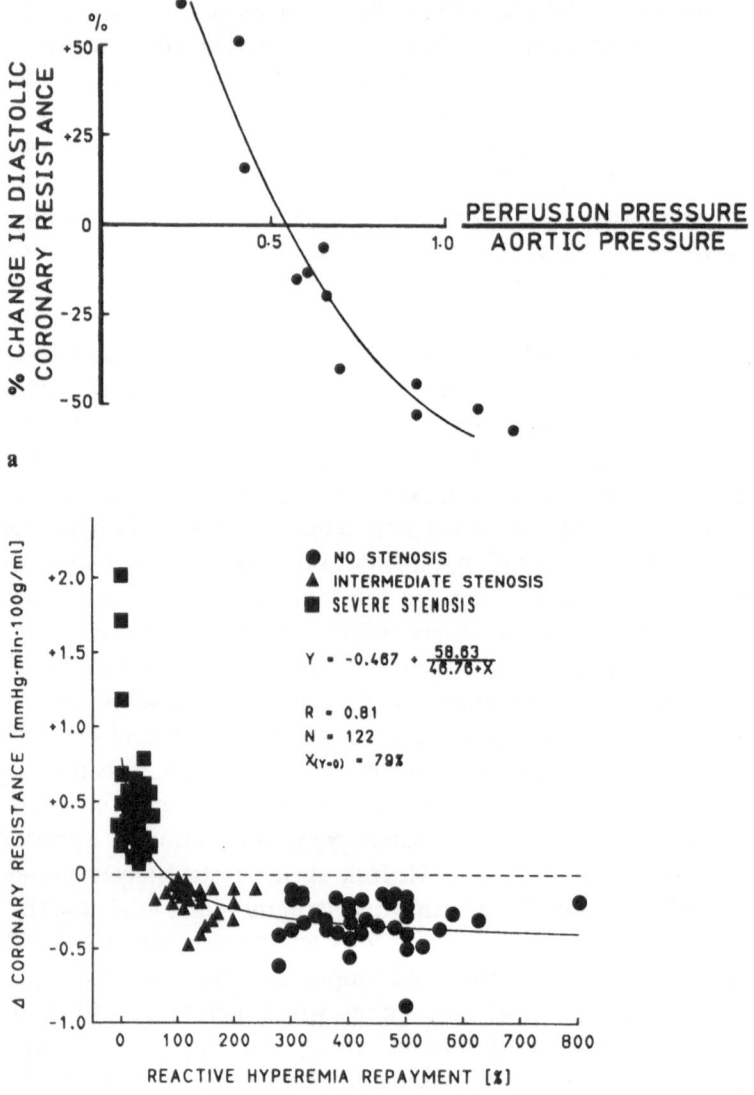

Fig. 10. a Changes in coronary resistance in response to intracoronary norepinephrine are plotted versus coronary perfusion pressure as a fraction of aortic pressure. From Bassenge et al. (1967). **b** Changes in coronary resistance in response to cardiac sympathetic nerve stimulation are plotted versus reactive hyperemia repayment as a measure of coronary dilator reserve. From Heusch and Deussen (1983) by permission of the American Heart Association. In both studies there is a progressive shift from norepinephrine-induced coronary dilation to coronary constriction with progressive coronary hypoperfusion.

further pursued until Buffington and Feigl (1981) demonstrated the persistence of α-adrenergic coronary vasoconstriction distal to a moderate stenosis during intracoronary norepinephrine infusion. This poststenotic vasoconstriction was powerful enough to limit oxygen supply up to the point of cardiac failure. However, net lactate production did not occur.

During electrical cardiac sympathetic nerve stimulation in anesthetized dogs, metabolic coronary dilation was progressively reversed to α-adrenergic coronary constriction when coronary dilator reserve was progressively reduced by an increaseing severity of coronary stenoses (Fig. 10b; Heusch and Deussen 1983). Distal to severe coronary stenoses, which exhausted poststenotic coronary dilator reserve, significant α-adrenergic coronary constriction was induced by cardiac sympathetic nerve stimulation, resulting in the precipitation of myocardial ischemia as evidenced by regional contractile dysfunction and net lactate production. The poststenotic coronary constriction and resulting ischemia were prevented by phentolamine and selective α_2-blockade with rauwolscine in the absence and presence of β-blockade (Heusch and Deussen 1983). Nifedipine also functionally antagonized the α_2-adrenoceptor mediated poststenotic coronary constriction and prevented the resulting myocardial ischemia (Heusch and Deussen 1984). Recently, an enhanced α_2-adrenergic constrictor tone in hypoperfused canine myocardium was confirmed by Chilian and Layne, using intravital microscopy (Chilian and Layne, 1990b).

The presence of α-adrenergic constrictor tone during coronary hypoperfusion was confirmed in anesthetized open-chest dogs with a presumably high resting sympathetic tone (Liang and Jones 1985; Jones et al. 1986). This α-adrenergic constrictor tone was attenuated by the selective α_1-antagonist prazosin and not by the selective α_2-antagonist yohimbine (Liang and Jones 1985). The different involvement of α_1- (Liang and Jones 1985) or α_2-adrenoceptors (Heusch and Deussen 1983) in coronary vasoconstriction during myocardial ischemia may be related to the different level of sympathetic activation in these studies.

In contrast to all previously discussed studies which agree on a deleterious role of α-adrenergic coronary constriction during myocardial ischemia, Nathan and Feigl (1986) concluded from their data that α-adrenergic coronary constriction exerts a favorable effect on ischemic myocardium by preventing a transmural redistribution of blood flow away from the ischemic subendocardium. In anesthetized dogs they observed, during constant inflow coronary hypoperfusion, an increase in interpolated subendocardial to subepicardial blood flow ratios in a control versus a phenoxybenzamine-treated region during intracoronary norepinephrine infusion. However, the use of constant inflow coronary hypoperfusion prevented the most significant beneficial effect of α-blockade in the studies described above, i. e., an increase in total coronary blood flow, and absolute data for subendocardial blood flow or measures of regional contractile function and metabolism were not presented to substantiate the aggravation of ischemia in the phenoxybenzamine-treated region. Under ischemic circumstances, the complete α_1- but incomplete α_2-blockade by phenoxybenzamine (Constantine and Lebel 1980) is particularly problematic since α_1-adrenergic coronary constriction is attenuated by ischemia (Heusch et al. 1983) whereas α_2-adrenergic coronary constriction is not (Deussen et al. 1985). Thus, phenoxybenzamine may have removed α_1-constrictor tone in the less ischemic subepicardium and left α_2-constrictor tone in the most ischemic subendocardium unopposed, causing an artificial imbalance of transmural blood flow distribution rather than proving a protective role of physiological α-adrenergic coronary constriction.

Transmurally nonuniform α_2-adrenergic coronary vasoconstriction contributing to the severity of myocardial ischemia has indeed been demonstrated in conscious, β-blocked dogs when, during the sympathetic activation by exercise, ischemia was induced by an acute coronary stenosis (Fig. 11; Seitelberger et al. 1988). Intracoronary infusion of the selective α_2-antagonist idazoxan distal to the stenosis resulted in significant increases in ischemic subendocardial and midmyocardial blood flow whereas subepicardial blood flow was unchanged (Fig. 12). Whereas the transmural

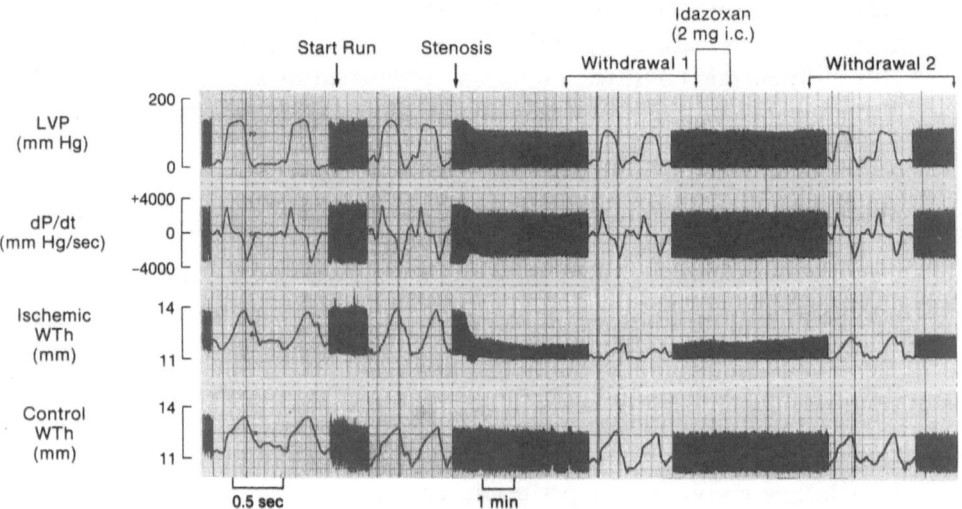

Fig. 11. Original recording demonstrating the improvement of ischemic regional myocardial wall thickening in a conscious dog during treadmill exercise by intracoronary infusion of the selective α_2-antagonist idazoxan distal to the coronary stenosis. From Seitelberger et al. (1988) by permission of the American Heart Association.

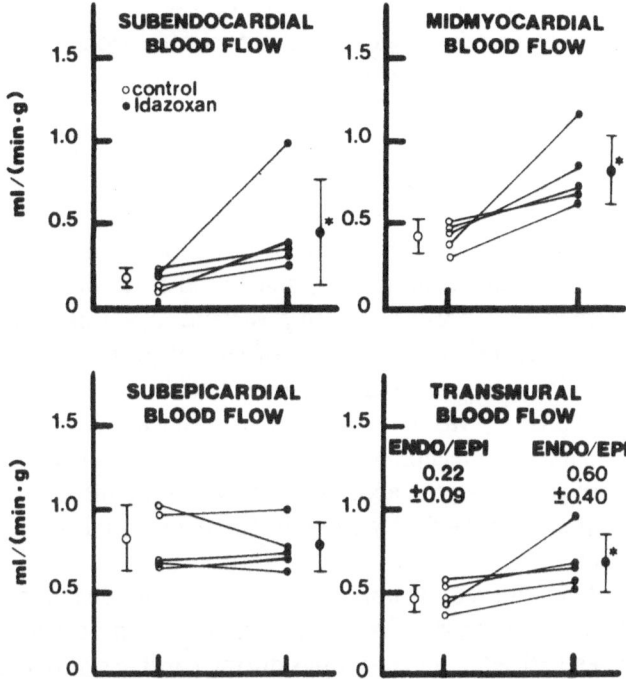

Fig. 12. Effects of the selective α_2-antagonist idazoxan on regional myocardial blood flow in conscious dogs with a severe coronary stenosis inflated during exercise. Idazoxan significantly improves blood flow to the most ischemic subendo- and midmyocardium whereas blood flow to the less ischemic subepicardium remains unchanged. From Seitelberger et al. (1988) by permission of the American Heart Association.

distribution of coronary vascular α_1- and α_2-adrenoceptors is uniform in nonischemic myocardium (Chen et al. 1988), the preferential α_2-adrenergic coronary constriction in the most ischemic inner myocardial layers (Seitelberger et al. 1988) may be related to the attenuation of α_1- but not α_2-adrenergic coronary constriction by ischemia (Deussen et al. 1985; Heusch et al. 1983), potentially due to the different sensitivity of α_1- and α_2-adrenoceptors to acidosis (McGrath 1982). In a very recent study in conscious dogs with constant pressure coronary hypoperfusion during exercise, intracoronary infusion of the selective α_1-antagonist prazosin significantly improved regional myocardial blood flow whereas selective α_2-blockade by intracoronary idazoxan caused only an insignificant increase in ischemic myocardial blood flow (Laxson et al. 1989). However, there are some methodological problems with this study since (1) it was conducted in the absence of β-blockade, (2) ischemia was more severe in the run with α_2-blockade, and (3) the dose of idazoxan was very low.

Recruitment of coronary dilator reserve, presumably through attenuation of α-adrenergic coronary constriction by the Ca^{2+} antagonist nifedipine, improved subendocardial and midmyocardial blood flow and attenuated regional contractile dysfunction during exercise-induced ischemia in conscious dogs (Heusch et al. 1987).

A true potective role of cardiac sympathetic nerves in an ischemic myocardial region through preferential subepicardial α-adrenergic coronary constriction was demonstrated in conscious dogs during exercise-induced ischemia, since subendocardial blood flow was higher in the innervated than in a phenol-denervated poststenotic region (Chilian and Ackell 1988). However, this protective role of subepicardial α-adrenergic coronary constriction was strictly limited to neuronally released norepinephrine whereas the coronary vasomotor effects in exercising dogs are essentially exerted by circulating catecholamines (Chilian et al. 1986). Thus, systemic α-blockade with phentolamine actually increased ischemic subendocardial blood flow in the innervated as well as in the denervated region (Chilian and Ackell 1988).

The failure to demonstrate any α-adrenergic coronary constriction distal to a severe coronary stenosis with exhausted poststenotic dilator reserve in anesthetized open-chest dogs during cardiac sympathetic nerve stimulation (Kopia et al. 1986) was probably due to the use of pentobarbital anesthesia, which is known to selectively inhibit α-adrenergic coronary vasoconstriction (Gwirtz and Stone 1982; Rosendorff et al. 1981). The relatively weak poststenotic coronary vasoconstriction in pigs during intravenous norepinephrine infusion, which did not result in a reduction of regional myocardial blood flow (Gewirtz et al. 1982), was probably related to the paucity of α-adrenoceptors in the coronary circulation of pigs (Schulz et al. 1990).

In conclusion, α-adrenergic constrictor tone in ischemic myocardium can be demonstrated in appropriate experimental models. α-Adrenergic coronary vasoconstriction acts to initiate or aggravate poststenotic myocardial ischemia. The involvement of α_1- and α_2-adrenoceptors is not clear yet.

5.3.3 α-Adrenergic Coronary Constriction in Clinical Myocardial Ischemia

5.3.3.1 α-Adrenoceptor-Induced Spasm

There are several more or less anecdotical observations of α-adrenoceptor involvement in coronary vasospasm (Levene and Freeman 1976; Raizner et al. 1980; Tzivoni et al. 1983; Yasue et al. 1976) which may be precipitated by the combined administration of epinephrine and propranolol (Yasue et al. 1976) or the cold pressor test (Raizner et al. 1980) and can be prevented by phenoxybenzamine (Levene and Freeman 1976; Yasue et al. 1976) or prazosin (Tzivoni et al. 1983). However, carefully controlled clinical trials have ruled out the possibility that α-adrenergic coronary constriction plays a significant role in the precipitation of vasospasm in epicardial coronary arteries (Chierchia et al. 1984; Robertson et al. 1983; Winniford et al. 1983).

5.3.3.2 Dynamic Coronary Stenosis

In a stenotic coronary arterial segment with eccentric atherosclerosis – with other parts of the wall retaining vasomotion – sympathetic activation by isometric exercise can induce critical narrowing (Brown et al. 1984a; Hossack et al. 1984), which then results in ischemic myocardial dysfunction and angina pectoris (Brown et al. 1984a). This critical narrowing of an epicardial coronary segment with preexisting eccentric atherosclerosis is presumably due to α-adrenoceptor activation during isometric exercise (Brown et al. 1984b). It can be prevented by intracoronary nitroglycerin (Brown et al. 1984a) or the Ca^{2+} antagonist diltiazem in a dose that does not induce unspecific epicardial coronary dilation (Hossack et al. 1984). Vasoconstriction of stenotic coronary arteries resulting in myocardial ischemia is also induced by dynamic exercise in patients with chronic stable angina, and again prevented by intracoronary nitroglycerin (Gage et al. 1986).

5.3.3.3 Cold Pressor Test

The cold pressor test is used as a provocative intervention to study the effects of reflex sympathetic activation in patients with coronary artery disease (Mueller et al. 1982). Whereas intact coronary arterial segments dilate during the cold pressor test, there is a preferential vasoconstriction of atherosclerotic coronary arterial segments (Nabel et al. 1988a; Zeiher et al. 1989b). A significant increase in coronary vascular resistance is induced by the cold pressor test in patients with coronary artery disease, resulting in the precipitation of angina pectoris in some of them (Mudge et al. 1976, 1979; Zeiher et al. 1989b). This increase in coronary resistance appears to be mediated by activation of α-adrenoceptors since it is prevented by phentolamine (Mudge et al. 1976). Nifedipine prevents both the increase in coronary vascular resistance and the net lactate production indicative of myocardial ischemia during the cold pressor test (Gunther et al. 1981). However, caution should be used in interpreting these studies because they were performed either using angiography of epicardial coronary segments (Nabel et al. 1988a), which may contribute little to coronary resistance (see above), or using coronary sinus thermodilution techniques (Gunther et al. 1981; Mudge et al. 1976) and Doppler flow velocity measurements (Zeiher et al. 1989b), which may be unreliable in the presence of flow heterogeneities caused by coronary artery disease. Additional studies on this subject are certainly needed.

5.3.3.4 Effort Angina

There is unequivocal evidence that α-adrenergic coronary vasoconstriction plays a significant role in exercise-induced myocardial ischemia in patients with stable angina pectoris (Berkenboom et al. 1986; Chierchia et al. 1985; Collins and Sheridan 1985; Gould et al. 1973). The exercise-induced ST-segment depression and angina pectoris in patients with chronic stable angina are reduced by intracoronary phentolamine (Fig. 13; Berkenboom et al. 1986; Chierchia et al. 1985), and exercise duration is prolonged by intravenous phentolamine (Gould et al. 1973). A reduction in exercise-induced ST-segment depression and an increase in exercise capacity are also achieved in patients with chronic stable angina treated with the selective α_1-antagonist indoramin (Collins and Sheridan 1985). However, in a recent study the exercise-induced ST-segment depression in patients with chronic stable angina was significantly more attenuated by the nonselective α-antagonist phentolamine than by the selective α_1-antagonist indoramin, suggesting that vascular α_2-adrenoceptors play an important role in the human coronary circulation (Berkenboom and Unger 1990).

5.4 Noncholinergic, Nonadrenergic Changes in Coronary Vasomotor Tone

In addition to the classical, more or less well analyzed neurotransmitters acetylcholine and norepinephrine, cardiac nerves contain peptides (neuropeptide tyrosine, NPY; calcitonin gene-related peptide, CGRP; vasoactive intestinal peptide, VIP; substance P, SP) which exert effects on coronary blood flow when administered exogenously.

NPY is found in human cardiac nerves with a particularly high concentration around coronary vessels (Gu et al. 1983). In the hearts of guinea pigs and pigs, NPY is coreleased with norepinephrine during cardiac sympathetic nerve stimulation (Haass et al. 1989; Rudehill et al. 1986). Intracoronary infusion of NPY induces a profound coronary vasoconstriction in guinea pigs, dogs, and pigs which is resistant to α-blockade (Aizawa et al. 1985; Franco-Cereceda et al. 1985; Rudehill et al. 1986) but antagonized by nifedipine (Franco-Cereceda et al. 1985; Rudehill et al. 1986) and

Fig. 13. Significant attenuation of exercise-induced ischemic ST-segment depression by intracoronary phentolamine (*closed circles*) at heart rates identical to those during a control exercise protocol (*open circles*) in patients with chronic stable angina. From Berkenboom et al. (1986) by permission.

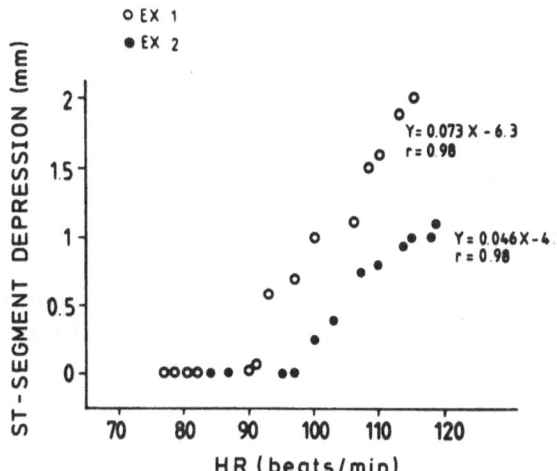

cyclooxygenase inhibitors (Martin and Patterson 1989). In patients with a history of angina pectoris, intracoronary infusion of NPY induces myocardial ischemia as evidenced by typical pain and ECG changes. Coronary angiography suggests a profound vasoconstriction of the distal vascular bed with no epicardial vasoconstriction (Clarke et al. 1987). In isolated rabbit coronary arteries, NPY potentiates norepinephrine-induced contractions and inhibits vasodilator-induced relaxations (Han and Abel 1987). The potentiation of adrenergic vasoconstriction in rabbit ear arteries is dependent on the integrity of the endothelium (Daly and Hieble 1987).

CGRP is found in cardiac nerves and its intracoronary infusion induces profound coronary vasodilation in pigs and humans (Ezra et al. 1987; McEwan et al. 1986). CGRP partially relieves the ergonovine-induced spasm in patients with variant angina (McEwan et al. 1986). VIP is also present in cardiac nerve fibers and induces both direct coronary vasodilation and metabolic coronary dilation secondary to an increase in myocardial performance (Anderson et al. 1988 a, b; Smitherman et al. 1989). Whereas substance P increases coronary blood flow in pigs (Ezra et al. 1986), probably by an endothelium-mediated mechanism, it induces vasoconstriction in isolated rat hearts (Kulakowski et al. 1983).

ATP may be coreleased with norepinephrine from cardiac sympathetic nerves reaching the vascular smooth muscle cells through the adventitial route. Depending on the type of coronary vascular purinergic receptor, ATP induces constriction or dilation of the coronary vessels in isolated rat hearts (Hopwood and Burnstock 1987).

While the above mentioned substances are certainly present in cardiac nerves and exert marked effects on coronary vasomotor tone, both their role as transmitters and their functional importance in physiological rather than pharmacological concentrations remain to be established.

6 Interaction of Different Control Mechanisms and Pathophysiological Implications

The interaction of different control mechanisms in the coronary circulation is extremely complex and dynamic. As mentioned above, elimination of one control mechanism can induce a compensatory increase in the functional significance of other control mechanisms, thus prohibiting a simple conclusion to be drawn from such experiments.

Under physiological conditions, a critical reduction in coronary blood flow initiating myocardial ischemia is prevented predominantly by the powerful local metabolic control of coronary vasomotor tone in concert with the modulatory role of autacoids released from a functionally intact endothelium, myogenic mechanisms, and finally humoral or neuronal control mechanisms adjusting coronary blood flow during homeostatic regulatory processes. Under pathological conditions, primarily the endothelial integrity is affected and the endothelium loses its anti-platelet adhesion and aggregation properties as well as its dilator function, leaving the vasculature exposed to direct stimulation by various circulating constrictor compounds and to the release products of adhering and aggregating platelets (TxA_2, 5-HT, ATP/ADP). With increasing severity of a coronary stenosis, the local autoregulatory reserve of the poststenotic coronary vascular bed is progressively recruited and finally exhausted. It is under these conditions of impaired endothelial function and exhausted local autoregulatory reserve that physiologically minor influences exerted by neuronal or humoral transmitters may actually prevail, then superimpose critical changes in coronary vasomotor tone, and ultimately precipitate myocardial ischemia.

An interesting hypothesis on the pathophysiological interaction of endothelial damage and platelet 5-hydroxytryptamine with cardiac sympathetic nerves has been forwarded by Cohen (1988; Cohen et al. 1987). After catheter-induced intimal injury and endothelial damage in dogs, 5-hydroxytryptamine from platelets aggregating at the site of intimal damage may be accumulated in coronary sympathetic nerves as a false transmitter in exchange for norepinephrine. There, 5-hydroxytryptamine may be released during nerve stimulation and induce epicardial coronary vasoconstriction. This interesting experimental finding and hypothesis exemplifies the complex nature of physiological and, even more so, of pathophysiological changes in coronary vasomotor tone, since it involves intimal damage of the coronary vascular wall, platelet aggregation, and activation of cardiac sympathetic nerves.

Acknowledgements. We thank Brian D. Guth, Ph. D., for the thorough review of this manuscript, Priv. Doz. Dr. Andreas Deussen for his review of the adenosine section as well as Prof. Dr. Karsten Schrör for his review of the prostaglandin-thromboxane section. We thank Emily Guth for language editing. The secretarial help of Ms. Bethina Blank, Ms. Karin Großkop and Ms. Inge Meier is appreciated.
The authors' contributions were supported by the Deutsche Forschungsgemeinschaft (Ba 408/13 and He 1320/1,2,3).

References

Aizawa Y, Murata M, Hayashi M, Funazaki T, Ito S, Shibata A (1985) Vasoconstrictor effect of neuropeptide Y (NPY) on canine coronary artery. Jpn J Physiol 49:584–588

Amezcua JL, Palmer RMJ, De Souza BM, Moncada S (1989) Nitric oxide synthesized from L-arginine regulates vascular tone in the coronary circulation of the rabbit. Br J Pharmacol 97:1119–1124

Anderson FL, Kralios AC, Hershberger R, Bristow MR (1988a) Effect of vasoactive intestinal peptide on myocardial contractility and coronary blood flow in the dog: comparison with isoproterenol and forskolin. J Cardiovasc Pharmacol 12:365–371

Anderson FL, Kralios AC, Hershberger R, Bristow MR (1988b) Desensitization of myocardial but not coronary VIP receptor-mediated responses in dogs. Am J Physiol 255: H601–H607

Ando J, Kamatsuda T, Kamiya A (1988) Cytoplasmic calcium response to fluid shear stress in cultured vascular endothelial cells. In Vitro Cell Develop Biol 24:871

Angus JA, Cocks TM (1989) Endothelium-derived relaxing factor. Pharmacol Ther 41:303–351

Armour JA, Randall WC (1971) Canine left ventricular intramyocardial pressures. Am J Physiol 220:1833–1839

Arnold G, Kosche F, Miessner E, Neitzert A, Lochner W (1968) The importance of the perfusion pressure in the coronary arteries for the contractility and the oxygen consumption of the heart. Pfluegers Arch 299:339–356

Ashton JH, Benedict CR, FitzGerald C, Raheja S, Taylor A, Campbell WB, Buja LM, Willerson JT (1986) Serotonin as a mediator of cyclic flow variations in stenosed canine coronary arteries. Circulation 73:572–578

Aversano T, Becker LC (1985) Persistence of coronary vasodilator reserve despite functionally significant flow reduction. Am J Physiol 248:H403–H411

Bache RJ, Cobb FR (1977) Effect of maximal coronary vasodilation on transmural myocardial perfusion during tachycardia in the awake dog. Circ Res 41:648–653

Bache RJ, Schwartz JS (1982) Effect of perfusion pressure distal to a coronary stenosis on transmural myocardial blood flow. Circulation 65:928–935

Bache RJ, Cobb FR, Greenfield Jr JC (1974) Myocardial blood flow distribution during ischemia-induced coronary vasodilation in the unanesthetized dog. J Clin Invest 54:1462–1472

Bardenheuer H, Schrader J (1983) Relationship between myocardial oxygen consumption, coronary flow, and adenosine release in an improved isolated working heart preparation of guinea pigs. Circ Res 51:263–271

Bassenge E (1984) Physiologie der Koronardurchblutung. In: Roskamm H (ed) Handbuch der inneren Medizin. Koronarerkrankungen. Springer, Berlin Heidelberg New York, pp 1–48

Bassenge E (1989) Flow-dependent regulation of coronary vasomotor tone. Eur Heart J 10 (Suppl F):22–27

Bassenge E, Busse R (1988) Endothelial modulation of coronary tone. Prog Cardiovasc Dis 30:349–380

Bassenge E, Pohl U (1986) Two principles of large artery dilation: indirect endothelium-mediated and direct smooth muscle relaxation. In: Magro E, Osswald W, Reis O, Vanhoutte P (eds) Central and peripheral mechanisms of cardiovascular regulation. Nato Adv Sci Inst Series A: Life Sciences. Plenum, New York, pp 163–196

Bassenge E, Walter P, Doutheil U (1967) Wirkungsumkehr der adrenergischen Coronargefässreaktion in Abhängigkeit vom Coronargefässtonus. Pfluegers Arch 297:146–155

Bassenge E, Werle E, Holtz J, Fritz H (1969) Significance of kinins in the coronary circulation. Pharmacol Res Comm 1:136–137

Bassenge E, Werle E, Walter P, Holtz J (1970) Significance of kinins in the coronary circulation. Adv Exp Med Biol 8:141–148

Bassenge E, Kucharczyk M, von Restorff W, Werle E (1972) Effect of bradykinin potentiating peptide on coronary circulation in conscious dogs. Adv Exp Med Biol 21:251–157

Bassenge E, Busse R, Pohl U (1987) Abluminal release and asymmetrical response of the rabbit arterial wall to endothelium-derived relaxing factor. Circ Res 61 (Suppl II):II-68-II-73

Bassenge E, Busse R, Pohl U (1989) Hemmung der Thrombozytenaggregation und -adhäsion durch EDRF und deren pathophysiologische Bedeutung. Z Kardiol 78 (Suppl 6):54–58

Bayliss WM (1902) On the local reaction of the arterial wall to changes of internal pressure. J Physiol 28:220–231

Bellamy RF (1978) Diastolic coronary artery pressure-flow relations in the dog. Circ Res 43:92–101

Bellamy RF (1980) Calculation of coronary vascular resistance. Cardiovasc Res 14:261–269

Belloni FL, Bruttig SP, Rubio R, Berne RM (1986) Uptake and release of adenosine by cultured rat aortic smooth muscle. Microvasc Res 32:200–210

Berkenboom GM, Unger P (1990) Alpha-adrenergic coronary constriction in effort angina. Basic Res Cardiol (in press)

Berkenboom GM, Abramowicz M, Vandermoten P, Degre SG (1986) Role of alpha-adrenergic coronary tone in exercise-induced angina pectoris. Am J Cardiol 57:195–198

Berne RM (1958) Effect of epinephrine and norepinephrine on coronary circulation. Circ Res 6:644–655

Berne RM (1961) Nucleotide degradation in the hypoxic heart and its possible relation to regulation of coronary blood flow. Fed Proc 20:101 (abstract)

Berne RM (1963) Cardiac nucleotides in hypoxia: possible role in regulation of coronary blood flow. Am J Physiol 204:317–322

Berne RM (1980) The role of adenosine in the regulation of coronary blood flow. Circ Res 47:807–813

Berne RM, Rubio R (1979) Coronary circulation. In: Handbook of Physiology. The cardiovascular system I. American Physiological Society, Maryland, pp 873–952

Berne RM, Blackmon JR, Gardner TH (1957) Hypoxemia and coronary blood flow. J Clin Invest 36:1101–1106

Berne RM, DeGeest H, Levy MN (1965) Influence of the cardiac nerves on coronary resistance. Am J Physiol 208:763–769

Boeynaems JM (1988) Drugs influencing the vascular production of prostacyclin. Prostaglandins 34:197–204

Bonham AC, Gutterman DD, Arthur JM, Marcus ML, Gebhart GF, Brody MJ (1987a) Electrical stimulation in perifornical lateral hypothalamus decreases coronary blood flow in cats. Am J Physiol 252:H474–H484

Bonham AC, Gutterman DD, Arthur JM, Marcus ML, Gebhart GF, Brody MJ (1987b) Neurogenic regulation of coronary blood flow: evidence for a central nervous system pathway. Circ Res 61 (Suppl II):II-42-II-46

Borchard F (1978) The adrenergic nerves of the normal and the hypertrophied heart. Thieme, Stuttgart

Bossaller C, Yamamoto H, Henry PD (1986) Endothelium-dependent relaxation is depressed in atherosclerotic arteries in vivo. Z Kardiol 75 (Suppl 4):63 (abstract)

Bossaller C, Habib GB, Yamamoto H, Williams C, Wells S, Henry PD (1987) Impaired muscarinic endothelium-dependent relaxation and cyclic guanosine 5'-monophosphate formation in atherosclerotic human coronary artery and rabbit aorta. J Clin Invest 79:170–174

Boulanger C, Lüscher TF (1990) Release of endothelin from the porcine aorta. Inhibition by endothelium-derived nitric oxide. J Clin Invest 85:587–590

Bretschneider HJ (1967) Aktuelle Probleme der Koronardurchblutung und des Myokardstoffwechsels. Regensburger Jbl Aerztl Fortb 15:1–27

Broderick R, Bialecki R, Tulenko TN (1989) Cholesterol-induced changes in rabbit arterial smooth muscle sensitivity to adrenergic stimulation. Am J Physiol 257:H170–H178

Brown BG, Lee AB, Bolson EL, Dodge HT (1984a) Reflex constriction of significant coronary stenosis as a mechanism contributing to ischemic left ventricular dysfunction during isometric exercise. Circulation 70:18–24

Brown BG, Bolson EL, Dodge HT (1984b) Dynamic mechanisms in human coronary stenosis. Circulation 70:917–922

Brum JM, Sufan Q, Dewey J, Bove AA (1985) Effects of angiotensin and ergonovine on large and small coronary arteries in the intact dog. Basic Res Cardiol 80:333–342

Buck JD, Hardman HF, Warltier DC, Gross GJ (1981) Changes in ischemic blood flow distribution and dynamic severity of a coronary stenosis induced by beta blockade in the canine heart. Circulation 64:708–715

Buckberg GD, Fixler DE, Archie Jr JP, Hoffman JIE (1972) Experimental subendocardial ischemia in dogs with normal coronary arteries. Circ Res 30:67–81

Buffington CW, Feigl EO (1981) Adrenergic coronary vasoconstriction in the presence of coronary stenosis in the dog. Circ Res 48:416–423

Bult H, Boeckxstaens GE, Pelckmans PA, Jordaens FH, van Maercke TM, Herman AG (1990) Nitric oxide as an inhibitory non-adrenergic non-cholinergic neurotransmitter. Nature (in press)

Burnstock G (1989) Vascular control by purines with emphasis on the coronary system. Eur Heart J 10 (Suppl F):15–21

Busch P, Deussen A, Heusch G (1988) Sympathetic effects on segmental coronary resistances and their role in coronary collateral perfusion. J Appl Cardiol 3:145–160

Buss DD, Wüsten B, Schaper W (1978) Effects of coronary stenoses and ventricular loading conditions on coronary flow. Basic Res Cardiol 73:571–583

Busse R, Mülsch A (1990) Calcium-dependent nitric oxide synthesis in endothelial cytosol is mediated by calmodulin. FEBS 265:133–136

Busse R, Trogisch G, Bassenge E (1985) The role of endothelium in the control of vascular tone. Basic Res Cardiol 80:475–490

Busse R, Lückhoff A, Bassenge E (1987) Endothelium-derived relaxant factor inhibits platelet activation. Naunyn Schmiedebergs Arch Pharmacol 336:566–571

Cambridge D, Davey MJ, Massingham R (1977) Prazosin, a selective antagonist of postsynaptic α-adrenoceptors. Br J Pharmacol 59:514P–515P

Canty JM (1988) Coronary pressure-function and steady-state pressure-flow relations during autoregulation in the unanesthetized dog. Circ Res 63:821–836

Canty JM, Klocke FJ (1985) Reduced regional myocardial perfusion in the presence of pharmacologic vasodilator reserve. Circulation 71:370–377

Carbonell LF, Carretero OA, Stewart JM, Scicli AG (1988) Effect of a kinin antagonist on the acute antihypertensive activity of enalaprilat in severe hypertension. Hypertension 11:239–243

Carlsson L, Abrahamsson T (1989) Ramiprilat attenuates the local release of noradrenaline in the ischemic myocardium. Eur J Pharmacol 166:157–164

Case RB, Greenberg H (1976) The response of canine coronary vascular resistance to local alterations in coronary arterial pCO_2. Circ Res 39:558–566

Case RB, Felix A, Wachter M, Kyriakidis G, Castellana F (1978) Relative effect of CO_2 on canine coronary vascular resistance. Circ Res 42:410–418

Cavero I, Dennis T, Lefevre-Borg F, Perrot P, Roach AG, Scatton B (1979) Effects of clonidine, prazosin and phentolamine on heart rate and coronary sinus catecholamine concentration during cardioaccelerator nerve stimulation in spinal dogs. Br J Pharmacol 67:283–292

Cernacek P, Stewart DJ (1989) Immunoreactive endothelin in human plasma: marked elevations in patients in cardiogenic shock. Biochem Biophys Res Comm 161:562–567

Chen DG, Dai X-Z, Zimmerman BG, Bache RJ (1988) Postsynaptic α_1- and α_2-adrenergic mechanisms in coronary vasoconstriction. J Cardiovasc Pharmacol 11:61–67

Chester A, Dashwood MR, Clarke J, Larkin S, Davies GJ, Tadjkarmi S, Maseri A, Yacoub M (1989) Influence of endothelin on human coronary arteries and localization of its binding sites. Am J Cardiol 63:1395–1398

Chierchia S, De Caterina R, Brunelli C, Crea F, Patrono C, Maseri A (1982) Failure of thromboxane A_2 blockade to prevent attacks of vasospastic angina. Circulation 62:702–705

Chierchia S, Davies G, Berkenboom G, Crea F, Crean P, Maseri A (1984) α-adrenergic receptors and coronary spasm: an elusive link. Circulation 69:8–14

Chierchia S, Pratt T, DeCoster P, Maseri A (1985) Alpha-adrenergic control of collateral flow: another determinant of coronary flow reserve. Circulation 72 (Suppl III):190 (abstract)

Chilian WM, Ackell PH (1988) Transmural differences in sympathetic coronary constriction during exercise in the presence of coronary stenosis. Circ Res 62:216–225

Chilian WM, Layne SM (1990a) Coronary microvascular responses to reductions in perfusion pressure. Evidence for persistent arteriolar vasomotor tone during coronary hypoperfusion. Circ Res 66:1227–1238

Chilian WM, Layne SM (1990b) Coronary arteriolar α_2-adrenergic constriction. FASEB J 4:A 1070 (abstract)

Chilian WM, Marcus ML (1985) Effects of coronary and extravascular pressure on intramyocardial and epicardial blood velocity. Am J Physiol 248:H170–H178

Chilian WM, Boatwright RB, Shoji T, Griggs DM (1981) Evidence against significant resting sympathetic coronary vasoconstrictor tone in the conscious dog. Circ Res 49:866–876

Chilian WM, Harrison DG, Haws CW, Snyder WD, Marcus ML (1986) Adrenergic coronary tone during submaximal exercise in the dog is produced by circulating catecholamines. Evidence for adrenergic denervation supersensitivity in the myocardium but not in coronary vessels. Circ Res 58:68–82

Chilian WM, Eastham CL, Layne SM, Marcus ML (1988) Small vessel phenomena in the coronary microcirculation: phasic intramyocardial perfusion and coronary microvascular dynamics. Prog Cardiovasc Dis 31:17–38

Chilian WM, Layne SM, Eastham CL, Marcus ML (1989) Heterogeneous microvascular coronary α-adrenergic vasoconstriction. Circ Res 64:376–388

Chu A, Stakely A, Cobb FR (1987) Nitrate-like effects of atrial natriuretic peptide on the coronary vasculature. Circulation 76 (Suppl IV):IV–240 (abstract)

Chu A, Cobb FR, Hagen PO, Murray JJ (1989a) Effects of a stabilized endothelium-derived relaxing factor on coronary vasculature in awake dogs. Am J Physiol 257:H1895–H1899

Chu A, Chambers D, Lin C-C, Kuehl W, Cobb FR (1989b) Nitric oxide significantly alters basal coronary vasomotor tone in the awake dog. Circulation 80 (Suppl II):II−124 (abstract)

Chu A, Morris K, Kuehl W, Cusma J, Navetta F, Cobb FR (1989c) Effects of atrial natriuretic peptide on the coronary arterial vasculature in humans. Circulation 80:1627−1635

Clarke JG, Kerwin R, Larkin S, Lee Y, Yacoub M, Davies GJ, Hackett D, Dawbarn D, Bloom SR, Maseri A (1987) Coronary artery infusion of neuropeptide Y in patients with angina pectoris. Lancet I:1057−1059

Clarke JG, Benjamin N, Larkin SW, Webb DJ, Keogh BE, Davies GJ, Maseri A (1989) Endothelin is a potent long-lasting vasoconstrictor in man. Am J Physiol 257:H2033−H2035

Clozel J-P, Clozel M (1989) Effects of endothelin on the coronary vascular bed in open-chest dogs. Circ Res 65:1193−1200

Cocks TM, Angus JA (1983) Endothelium-dependent relaxation of coronary arteries by noradrenaline and serotonin. Nature 305:627−629

Cocks TM, Angus JA, Campbell GR (1985) Release and properties of endothelium-derived relaxing factor (EDRF) from endothelial cells in culture. J Cell Physiol 123:310−320

Cohen MV, Kirk ES (1973) Differential response of large and small coronary arteries to nitroglycerin and angiotensin. Circ Res 33:445−453

Cohen RA (1986) Role of autonomic nerves and endothelium in coronary vasospasm. In: Tulenko T (ed) Recent advances in arterial diseases: atheroslerosis, hypertension, and vasospasm. Liss, New York, pp 353−362

Cohen RA (1988) Platelet 5-hydroxytryptamine and vascular adrenergic nerves. News Physiol Sci 3:185−189

Cohen RA, Cunningham LD (1988) Low density lipoproteins inhibit endothelium-dependent relaxations caused by bradykinin in the pig coronary artery. Circulation 78 (Suppl II):II−183 (abstract)

Cohen RA, Weisbrod RM (1988) Endothelium inhibits norepinephrine release from adrenergic nerves of rabbit carotid artery. Am J Physiol 254:H871−H878

Cohen RA, Shepherd JT, Vanhoutte PM (1983a) Inhibitory role of the endothelium in the response of isolated coronary arteries to platelets. Science 221:273−274

Cohen RA, Shepherd JT, Vanhoutte PM (1983b) Prejunctional and postjunctional actions of endogenous norepinephrine at the sympathetic neuroeffector junction in canine coronary arteries. Circ Res 52:16−25

Cohen RA, Shepherd JT, Vanhoutte PM (1984) Effects of the adrenergic transmitter on epicardial coronary arteries. Fed Proc 43:2862−2870

Cohen RA, Zitnay KM, Weisbrod RM (1987) Accumulation of 5-hydroxytryptamine leads to dysfunction of adrenergic nerves in canine coronary artery following intimal damage in vivo. Circ Res 61:829−833

Cohen RA, Zitnay KM, Haudenschild CC, Cunningham LD (1988) Loss of selective endothelial cell vasoactive functions caused by hypercholesterolemia in pig coronary arteries. Circ Res 63:903−910

Collier J, Vallance P (1989) Endothelium-derived relaxing factor is an endogenous vasodilator in man. Br J Pharmacol 97:639−641

Collins P, Sheridan D (1985) Improvement in angina pectoris with alpha adrenoceptor blockade. Br Heart J 53:488−492

Connaughton S, Docherty JR (1988) Evidence that SK and F 104078 does not differentiate between pre- and postjunctional α_2-adrenoceptors. Naunyn Schmiedebergs Arch 338:379−382

Connaughton S, Docherty JR (1990) No evidence for differences in pre- and postjunctional α_2-adrenoceptors in the periphery. Br J Pharmacol 99:97−102

Constantine JW, Lebel W (1980) Complete blockade by phenoxybenzamine of alpha 1- but not of alpha 2- vascular receptors in dogs and the effect of propranolol. Naunyn Schmiedebergs Arch Pharmacol 314:149−156

Constantine JW, Weeks RA, McShane WK (1978) Prazosin and presynaptic α-receptors in the cardioaccelerator nerve of the dog. Eur J Pharmacol 50:51−60

Cooke JP, Rossitch Jr E, Anden E, Loscalzo J, Dzau VJ (1990) Flow activates a specific endothelial potassium channel to release an endogenous nitrovasodilator. J Clin Invest (in press)

Cox DA, Hintze TH, Vatner SF (1983) Effects of acetylcholine on large and small coronary arteries in conscious dogs. J Pharmacol Exp Ther 225:764−769

Cross CE, Rieben PA, Salisbury PF (1961) Influence of coronary perfusion and myocardial edema on pressure-volume diagram of left ventricle. Am J Physiol 201:102−108

Crossman DC, Larkin SW, Fuller RW, Davies GJ, Maseri A (1989) Substance P dilates epicardial coronary arteries and increases coronary blood flow in humans. Circulation 80:475−484

Dai X-Z, Herzog CA, Schwartz JS, Bache RJ (1986) Coronary blood flow during exercise following nonselective and selective α_1-adrenergic blockade with indoramin. J Cardiovasc Pharmacol 8:574−581

Daly RN, Hieble JP (1987) Neuropeptide Y modulates adrenergic neurotransmission by an endothelium dependent mechansim. Eur J Pharmacol 138:445−446

Daniell HB, Carson RR, Ballard KD, Thomas GR, Privitera PJ (1984) Effects of captopril on limiting infarct size in conscious dogs. J Cardiovasc Pharmacol 6:1043−1047

Dart AM, Schömig A, Dietz R, Mayer E, Kübler W (1984) Release of endogenous catecholamines in the ischemic myocardium of the rat. Part B: Effect of sympathetic nerve stimulation. Circ Res 55:702−706

Daugherty A, Zweifel BS, Sobel BE, Schonfeld G (1988) Isolation of low density lipoprotein from atherosclerotic vascular tissue of Watanabe heritable hyperlipidemic rabbits. Arteriosclerosis 8:768−777

Daut J, Maier-Rudolph W, van Beckerath N, Mehrke G, Günther K, Goedel-Meinen L (1990) Hypoxic dilation of coronary arteries is mediated by ATP-sensitive potassium channels. Science 247:1341−1344

Decker N, Schwartz PJ (1985) Postjunctional $alpha_1$- and $alpha_2$-adrenoceptors in the coronaries of the perfused guinea-pig heart. J Pharmacol Exp Ther 232:251−257

Denn MJ, Stone HL (1976) Autonomic innervation of dog coronary arteries. J Appl Physiol 41:30−35

De Nucci G, Thomas R, D'Orleans-Juste P, Antunes E, Walder C, Warner TD, Vane JR (1988) Pressor effects of circulating endothelin are limited by its removal in the pulmonary circulation and by the release of prostacyclin and endothelium-derived relaxing factor. Proc Natl Acad Sci USA 85:9797−9800

Deussen A, Heusch G, Thämer V (1985) Alpha 2-adrenoceptor-mediated coronary vasoconstriction persists after exhaustion of coronary dilator reserve. Eur J Pharmacol 115:147−153

Deussen A, Möser G, Schrader J (1986) Contribution of coronary endothelial cells to cardiac adenosine production. Pfluegers Arch 406:608−614

Deussen A, Borst M, Kroll K, Schrader J (1988) Formation of S-adenosylhomocysteine in the heart. II: A sensitive index for regional myocardial underperfusion. Circ Res 63:250−261

Deussen A, Lloyd HGE, Schrader J (1989) Contribution of S-adenosylhomocysteine to cardiac adenosine formation. J Mol Cell Cardiol 21:773−782

De Witt DF, Wangler RD, Thompson CI, Sparks HVJr (1983) Phasic release of adenosine during steady state metabolic stimulation in the isolated guinea pig heart. Circ Res 53:636−643

Dole WP (1987) Autoregulation of the coronary circulation. Prog Cardiovasc Disc 29:293–323

Dole WP, Bishop VS (1982) Influence of autoregulation and capacitance on diastolic coronary artery pressure-flow relationships in the dog. Circ Res 51:261–270

Dole WP, Nuno DW (1986) Myocardial oxygen tension determines the degree and pressure range of coronary autoregulation. Circ Res 59:202–215

Dole WP, Yamada N, Bishop VS, Olsson RA (1985) Role of adenosine in coronary blood flow regulation after reductions in perfusion pressure. Circ Res 56:517–524

Domenech RJ, de la Prida JM (1975) Mechanical effects of heart contraction on coronary flow. Cardiovasc Res 9:509–514

Downey HF, Bashour FA, Boatwright RB, Parker PE (1975) Uniformity of transmural perfusion in anesthetized dogs with maximally dilated coronary circulations. Circ Res 37:111–117

Downey JM, Kirk ES (1975) Inhibition of coronary blood flow by a vascular waterfall mechanism. Circ Res 36:753–760

Doxey JC, Roach AG, Smith CFC (1983) Studies on RX 781094: a selective, potent and specific antagonist of alpha 2-adrenoceptors. Br J Pharmacol 78:489–505

Drexler H, Zeiher AM, Wollschläger H, Meinertz T, Just H, Bonzel T (1989) Flow-dependent coronary artery dilatation in humans. Circulation 80:466–474

Driscol TE, Moir TW, Eckstein RW (1964) Vascular effect of changes in perfusion pressure in the nonischemic and ischemic heart. Circ Res 14/15 (Suppl I): I94–I102

Dull RO, Davies PF (1989) Featured Research: Endothelial regulation of vascular tone. Circulation 80 (Suppl II): II–481 (abstract)

Dusting GJ (1984) Coronary vasomotor tone: the role of prostanoids re-examined. Bibl Cardiol 38:178–188

Dusting GJ, Moncada S, Vane JR (1979) Prostaglandins, their intermediates and precursors: cardiovascular actions and regulatory roles in normal and abnormal circulatory systems. Prog Cardiovasc Dis 21:405–430

Dzau VJ (1988) Circulating versus local renin-angiotensin system in cardiovascular homeostasis. Circulation 77 (Suppl I):I4–I13

Edlund A, Berglund B, van Dorne D, Kaijser L, Nowak J, Patrono C, Sollevi A, Wennmalm A (1985) Coronary flow regulation in patients with ischemic heart disease: release of purines and prostacyclin and the effect of inhibitors of prostacyclin formation. Circulation 71:1113–1120

Ehring T, Heusch G (1990) Felodipine prevents the poststenotic myocardial ischemia induced by α_2-adrenergic coronary constriction. Cardiovasc Drugs Ther 4:443–450

Eikens E, Wilcken DEL (1974) Reactive hyperemia in the dog heart: effects of temporarily restricting arterial inflow and of coronary occlusions lasting one and two cardiac cycles. Circ Res 35:702–712

Ellis AK, Klocke FJ (1979) Effects of preload on the transmural distribution of perfusion and pressure-flow relationships in the canine coronary vascular bed. Circ Res 46:68–77

Ertl G (1987) Coronary vasoconstriction in experimental myocardial ischemia. J Cardiovasc Pharmacol 9 (Suppl 2):S9–S17

Ertl G, Kloner RA, Alexander RW, Braunwald E (1982) Limitation of experimental infarct size by an angiotensin-converting enzyme inhibitor. Circulation 65:40–48

Evers AS, Murphree S, Saffitz JE, Jakschik BA, Needleman P (1985) Effects of endogenously produced leukotrienes, thromboxane, and prostaglandins on coronary vascular resistance in rabbit myocardial infarction. J Clin Invest 75:992–999

Ezra D, Laurindo FRM, Eimerl J, Goldstein RE, Peck CC, Feuerstein G (1986) Tachykinin modulation of coronary blood flow. Eur J Pharmacol 122:135–138

Ezra D, Laurindo FRM, Goldstein DS, Goldstein RE, Feuerstein G (1987) Calcitonin gene-related peptide: a potent modulator of coronary flow. Eur J Pharmacol 137:101–105

Ezra D, Goldstein RE, Czaja JF, Feuerstein GZ (1989) Lethal ischemia due to intracoronary endothelin in pigs. Am J Physiol 257:H339–H343

Faber JE (1988) In situ analysis of α-adrenoceptors in arteriolar and venular smooth muscle in rat skeletal muscle microcirculation. Circ Res 62:37–50

Fam WM, McGregor M (1968) Effect of nitroglycerin and dipyridamole on regional coronary resistance. Circ Res 22:649–659

Fauler J, Frölich JC (1989) Cardiovascular effects of leukotrienes. Cardiovasc Drugs Ther 3:499–505

Feigl EO (1968) Carotid sinus reflex control of coronary blood flow. Circ Res 23:223–237

Feigl EO (1969) Parasympathetic control of coronary blood flow in dogs. Circ Res 25:509–519

Feigl EO (1975) Control of myocardial oxygen tension by sympathetic coronary vasoconstriction in the dog. Circ Res 37:88–95

Feigl EO (1983) Coronary physiology. Physiol Rev 63:1–205

Feigl EO (1987) The paradox of adrenergic coronary vasoconstriction. Circulation 76:737–745

Feldman RD, Christy JP, Paul ST, Harrison DG (1989) β-adrenergic receptors on canine coronary collateral vessels: characterization and function. Am J Physiol 257:H1634–H1639

Fiedler VB, Abram TS (1987) Effects of intracoronary leukotriene D4 infusion on the coronary circulation, hemodynamics, and prostanoid release in porcine experiments. Cardiology 74:89–99

Fiedler VB, Mardin M, Abram TS (1984) Leukotriene D4-induced vasoconstriction of coronary arteries in anaesthetized dogs. Eur Heart J 5:253–260

Fischell TA, Nellessen U, Johnson DE, Ginsburg R (1989) Endothelium-dependent arterial vasoconstriction after balloon angioplasty. Circulation 79:899–910

Fish RD, Nabel EG, Selwyn AP, Ludmer PL, Mudge GH, Kirshenbaum JM, Schoen FJ, Alexander RW, Ganz P (1988) Responses of coronary arteries of cardiac transplant patients to acetylcholine. J Clin Invest 81:21–31

Fitscha P, Kaliman J, Sinzinger H (1985) Gamma-camera imaging after autologous human platelet labeling with 111In-oxine-sulfate: a key for assessing the efficacy of prostacyclin treatment in active atherosclerosis? In: Schrör K (ed) Prostaglandins and other eicosanoids in the cardiovascular system. Karger, Basel, pp 352–357

FitzGerald GA, Smith B, Pederson AK, Brash AR (1984) Increased prostacyclin biosynthesis in patients with severe atherosclerosis and platelet activation. N Engl J Med 310:1065–1068

Fleckenstein A, Fleckenstein-Grün G (1988) Mechanism of action of calcium antagonists in heart and vascular smooth muscle. Eur Heart J 9 (Suppl H):95–99

Fleckenstein A, Nakayama K, Fleckenstein-Grün G, Byon YK (1975) Interactions of vasoactive ions and drugs with Ca-dependent excitation-contraction coupling of vascular smooth muscle. North Holland, Amsterdam, pp 555–566

Folkow B (1952) A study of the factors influencing the tone of denervated blood vessels perfused at various pressures. Acta Physiol Scand 27:99–117

Folkow B (1964) Description of the myogenic hypothesis. Circ Res 14/15 (Suppl I):I279–I287

Folts JD, Crowell Jr EB, Rowe GG (1976) Platelet aggregation in partially obstructed vessels and its elimination with aspirin. Circulation 54:365–370

Förster W (1980) Effect of various agents on prostaglandin biosynthesis and the anti-aggregatory effect. Acta Med Scand 642 [Suppl]6:35–46

Förstermann U, Mülsch A, Böhme E, Busse R (1986) Stimulation of soluble guanylate cyclase by an acetylcholine-induced endothelium-derived factor from rabbit and canine arteries. Circ Res 58:531–538

Förstermann U, Mügge A, Alheid U, Haverich A, Frölich JC (1988) Selective attenuation of endothelium-mediated vasodilation in atherosclerotic human coronary arteries. Circ Res 62:185–190

Franco-Cereceda A (1989) Endothelin- and neuropeptide Y-induced vasoconstriction of human epicardial coronary arteries. Br J Pharmacol 97:968–972

Franco-Cereceda A, Lundberg JM, Dahlöf C (1985) Neuropeptide Y and sympathetic control of heart contractility and coronary vascular tone. Acta Physiol Scand 124:361–369

Freiman PC, Mitchell GG, Heistad DD, Armstrong ML, Harrison DG (1986) Atherosclerosis impairs endothelium-dependent vascular relaxation to acetylcholine and thrombin in primates. Circ Res 58:783–789

Friedman PL, Brown EJ, Gunther S, Alexander RW, Barry WH, Mudge GH, Grossman W (1981) Coronary vasoconstrictor effect of indomethacin in patients with coronary artery disease. N Engl J Med 305:1171–1175

Fukuda K, Hori S, Kusuhara M, Satoh T, Kyotani S, Handa S, Nakamura Y, Oono H, Yamaguchi K (1989) Effect of endothelin as a coronary vasoconstrictor in the Langendorff-perfused rat heart. Eur J Pharmacol 165:301–304

Furchgott RF, Zawadzki JV (1980) The obligatory role of endothelial cells in the relaxation of arterial smooth muscle by acetylcholine. Nature 288:373–376

Furlong B, Hendersen AH, Lewis MJ, Smith JA (1987) Endothelium-derived relaxing factor inhibits in vitro platelet aggregation. Br J Pharmacol 90:687–692

Fuster V, Steele PM, Chesebro JH (1985) Role of platelets and thrombosis in coronary atherosclerotic disease and sudden death. J Am Coll Cardiol 5:175B–184B

Gage JE, Hess OM, Murakami T, Ritter M, Grimm J, Krayenbuehl HP (1986) Vasoconstriction of stenotic coronary arteries during dynamic exercise in patients with classic angina pectoris: reversibility by nitroglycerin. Circulation 73:865–876

Gallagher KP, Matsuzaki M, Osakada G, Kemper WS, Ross Jr J (1983) Effect of exercise on the relationship between myocardial blood flow and systolic wall thickening in dogs with acute coronary stenosis. Circ Res 52:716–729

Gallagher KP, Osakada G, Kemper WS, Ross Jr J (1985) Cyclical coronary flow reductions in conscious dogs equipped with ameroid constrictors to produce severe coronary narrowing. Basic Res Cardiol 80:100–106

Galle J, Bassenge E, Busse R (1990a) Oxidized low density lipoproteins potentiate vasoconstrictions to various agonists by direct interaction with vascular smooth muscle. Circ Res 66:1287–1293

Galle J, Lückhoff A, Busse R, Bassenge E (1990b) Effects of native and oxidized low density lipoproteins on formation and inactivation of EDRF. Arteriosclerosis (in press)

Gardiner S, Compton A, Bennet T, Palmer R, Moncada S (1990) Regional haemodynamic changes during oral ingestion of NG-monomethyl-L-arginine or NG-nitro-L-arginine methyl ester in conscious Brattleboro rats. Br J Pharmacol 99: (in press)

Garthwaite J, Charles SL, Chess-Williams R (1988) Endothelium-derived relaxing factor release on activation of NMDA receptors suggests role as intercellular messenger in the brain. Nature 336:385–388

Gavras H, Benetos A, Gavras I (1987) Contribution of bradykinin to maintenance of normal blood pressure. Hypertension 9 (Suppl III):III147–III149

Gebremedhin D, Koltei MZ, Pogatsa G, Magyer K, Hadhazy P (1988) Influence of experimental diabetes on the mechanical responses of canine coronary arteries: role of endothelium. Cardiovasc Res 22:537–544

Gerlach E, Deuticke B, Dreisbach RH (1963) Der Nucleotid-Abbau im Herzmuskel bei Sauerstoffmangel und seine mögliche Bedeutung für die Coronardurchblutung. Naturwissenschaften 6:228–229

Gerlings ED, Miller DT, Gilmore JP (1969) Oxygen availability: a determinant of myocardial potassium balance. Am J Physiol 216:559–562

Gerova M, Barta E, Gero J (1979a) Sympathetic control of major coronary artery diameter in the dog. Circ Res 44:459–467

Gerova M, Dolezel S, Gero J, Barta E (1979b) Role of the vagus in control of the major conduit coronary artery in the dog. Physiol Bohemoslov 28:299–307

Gerrity RG, Naito HK, Richardson M, Schwartz CJ (1979) Dietary induced atherogenesis in swine. Am J Pathol 95:775–792

Gewirtz H, Most AS, Williams DO (1982) The effect of generalized alpha-receptor stimulation on regional myocardial blood flow distal to a severe coronary artery stenosis. Circulation 65:1329–1336

Gewirtz H, Brautigan DL, Olsson RA, Brown P, Most AS (1983) Role of adenosine in the maintenance of coronary vasodilation distal to a severe coronary artery stenosis. Observations in conscious domestic swine. Circ Res 53:42–51

Gewirtz H, Weeks G, Nathanson M, Sharaf B, Fedele F, Most AS (1989) Tissue acidosis. Role in sustained arteriolar dilatation distal to a coronary stenosis. Circulation 79:890–898

Gibbs JSR, Crean PA, Mockus L, Wright C, Sutton GC, Fox KM (1989) The variable effect of angiotensin converting enzyme inhibition on myocardial ischaemia in chronic stable angina. Br Heart J 62: 112–117

Gibson A, Mirzazadeh S, Hobbs AJ, Moore PK (1990) L-NG-monomethyl-arginine and L-NG-nitro-arginine inhibit non-adrenergic, non-cholinergic relaxation of the mouse anococcygeus muscle. Br J Pharmacol 99:602–606

Ginsburg RC (1984) Myogenic tone of the isolated human epicardial artery: regulatory controls. Acta Med Scand 694 [Suppl]:29–37

Ginsburg R, Bristow MR, Davis K, Billingham ME, Schroeder JS, Stinson EB, Harrison DC (1981) Receptor analysis of the human coronary artery – normal distribution and effects of atherosclerosis. Circulation 64 [Suppl IV]:IV–120 (abstract)

Glazier JJ, Faxon DP, Mills RM, Bresnahan MR, Ryan TJ, Gavras H (1989) Effect of arginine vasopressin on coronary and systemic hemodynamics in man. Int J Cardiol 24:95–103

Gorman MW, Sparks HV Jr (1982) Progressive coronary vasoconstriction during relative ischemia in canine myocardium. Circ Res 51:411–420

Gould L, Reddy GV, Gombrecht RF (1973) Oral phentolamine in angina pectoris. Jpn Heart J 14:393–397

Grant JA, Scrutton MC (1979) Novel α_2-adrenoceptors primarily responsible for inducing human platelet aggregation. Nature 277:659–661

Greenberg S, Diecke FPJ, Peevy K, Tanaka RP (1989) The endothelium modulated adrenergic neurotransmission to canine pulmonary arteries and veins. Eur J Pharmacol 162:67–80

Gregg DE (1963) Effect of coronary perfusion pressure or coronary flow on oxygen usage of the myocardium. Circ Res 13:497–500

Griffith TM, Lewis MJ, Newby AC, Henderson AH (1988) Endothelium-derived relaxing factor. J Am Coll Cardiol 12:797–806

Gu J, Polak JM, Adrian TE, Allen JM, Tatemoto K, Bloom SR (1983) Neuropeptide tyrosine (NPY) – a major cardiac neuropeptide. Lancet I:1008–1010

Gunther S, Green L, Muller JE, Mudge GH, Grossman W (1981) Prevention by nifedipine of abnormal coronary vasoconstriction in patients with coronary artery disease. Circulation 63:849–855

Guth BD, Heusch G, Seitelberger R, Ross Jr J (1987a) Mechanism of beneficial effect of beta-adrenergic blockade on exercise-induced myocardial ischemia in conscious dogs. Circ Res 60:738–746

Guth BD, Heusch G, Seitelberger R, Ross Jr J (1987b) Elimination of exercise-induced regional myocardial dysfunction by a bradycardic agent in dogs with chronic coronary stenosis. Circulation 75:661–669

Guth BD, Thaulow E, Heusch G, Seitelberger R, Ross Jr J (1990) Myocardial effects of selective alpha-adrenoceptor blockade during exercise in dogs. Circ Res 66:1703–1712

Guz A, Kurland GS, Freedberg AS (1960) Relation of coronary flow to oxygen supply. Am J Physiol 199:179–182

Gwirtz PA, Stone HL (1982) Coronary blood flow changes following activation of adrenergic receptors in the conscious dog. Am J Physiol 243:H13–H19

Gwirtz PA, Overn SP, Mass HJ, Jones CE (1986) Alpha 1-adrenergic constriction limits coronary flow and cardiac function in running dogs. Am J Physiol 250:H1117–H1126

Haass M, Cheng B, Richardt G, Lang RE, Schömig A (1989) Characterization and presynaptic modulation of stimulation-evoked exocytotic co-release of noradrenaline and neuropeptide Y in guinea pig heart. Naunyn Schmiedebergs Arch Pharmacol 339:71–78

Hackett JG, Abboud FM, Mark AL, Schmid PG, Heistad DD (1972) Coronary vascular responses to stimulation of chemoreceptors and baroreceptors. Circ Res 31:8–17

Haddy FJ, Scott JB (1975) Metabolic factors in peripheral circulatory regulation. Fed Proc 34:2006–2011

Hamilton FN, Feigl EO (1976) Coronary vascular sympathetic beta-receptor innervation. Am J Physiol 230:1569–1576

Han C, Abel PW (1987) Neuropeptide Y potentiates contraction and inhibits relaxation of rabbit coronary arteries. J Cardiovasc Pharmacol 9:675–681

Harder DR (1987) Pressure-induced myogenic activation of cat cerebral arteries is dependent on intact endothelium. Circ Res 60:102–107

Harrison DG, Chilian WM, Marcus ML (1986) Absence of functioning alpha-adrenergic receptors in mature canine coronary collaterals. Circ Res 59:133–142

Harrison DG, Freiman PC, Armstrong ML, Marcus ML, Heistad DD (1987a) Alterations of vascular reactivity in atherosclerosis. Circ Res 61 (Suppl II):II74–II80

Harrison DG, Armstrong ML, Freiman PC, Heistad DD (1987b) Restoration of endothelium-dependent relaxation by dietary treatment of atherosclerosis. J Clin Invest 80:1808–1811

Hart G, Gokal R (1977) The syndrome of inappropriate antidiuretic hormone secretion associated with acute myocardial infarction. Postgrad Med J 53:761–762

Hartmann A, Saeed M, Bing RJ (1987) Release of endothelium-derived relaxing factor from freshly harvested porcine endothelial cells. Circ Res 61:548–554

Hautamaa PV, Dai XZ, Homans DC, Robb JF, Bache RJ (1987) Vasomotor properties of immature canine coronary collateral circulation. Am J Physiol 252:H1105–H1111

Hautamaa PV, Dai X-Z, Homans DC, Bache RJ (1989) Vasomotor activity of moderately well-developed canine coronary collateral circulation. Am J Physiol 256:H890–H897

Haws CW, Green LS, Burgess MJ, Abildskov JA (1987) Effects of cardiac sympathetic nerve stimulation on regional coronary blood flow. Am J Physiol 252:H269–H274

Heistad DD, Armstrong ML, Marcus ML, Piegors DJ, Mark AL (1984) Augmented responses to vasoconstrictor stimuli in hypercholesterolemic and atherosclerotic monkeys. Circ Res 54:711–718

Heistad DD, Mark AL, Marcus ML, Piegors DJ, Armstrong ML (1987) Dietary treatment of atherosclerosis abolishes hyperresponsiveness to serotonin: implications for vasospasm. Circ Res 61:346–351

Hennekens CH, Peto R, Hutchison GB, Doll R (1988) On overview of the British and American aspirin studies. N Engl J Med 318:923–924

Henry PD, Yokoyama M (1980) Supersensitivity of atherosclerotic rabbit aorta to ergonovine. J Clin Invest 66:306–313

Heusch G (1990) α-adrenergic mechanisms in myocardial ischemia. Circulation 81:1–13

Heusch G, Deussen A (1983) The effects of cardiac sympathetic nerve stimulation on the perfusion of stenotic coronary arteries in the dog. Circ Res 53:8–15

Heusch G, Deussen A (1984) Nifedipine prevents sympathetic vasoconstriction distal to severe coronary stenoses. J Cardiovasc Pharmacol 6:378–383

Heusch G, Guth BD (1989) Neurogenic regulation of coronary vasomotor tone. Eur Heart J 10 (Suppl F):6–14

Heusch G, Yoshimoto N (1983a) Effects of heart rate and perfusion pressure on segmental coronary resistances and collateral perfusion. Pfluegers Arch 397:284–289

Heusch G, Yoshimoto N (1983 b) Effects of cardiac contraction on segmental coronary resistances and collateral perfusion. Int J Microcirc 2:131 – 141

Heusch G, Yoshimoto N, Heegemann H, Thämer V (1983) Interaction of methoxamine with compensatory vasodilation distal to coronary stenoses. Drug Res 33:1647 – 1650

Heusch G, Deussen A, Schipke J, Thämer V (1984) α_1- and α_2-adrenoceptor-mediated vasoconstriction of large and small canine coronary arteries in vivo. J Cardiovasc Pharmacol 6:961 – 968

Heusch G, Deussen A, Schipke J, Thämer V (1986) Adenosine, dipyridamole and isosorbiddinitrate are ineffective to prevent the sympathetic initiation of poststenotic myocardial ischemia. Drug Res 36:1045 – 1048

Heusch G, Seitelberger R, Guth BD, Ross Jr J (1986) Adrenergic mechanisms in myocardial ischemia. J Appl Cardiol 1:125 – 142

Heusch G, Guth BD, Seitelberger R, Ross Jr J (1987) Attenuation of exercise-induced myocardial ischemia in dogs with recruitment of coronary vasodilator reserve by nifedipine. Circulation 75:482 – 490

Heyndrickx GR, Boettcher DH, Vatner SF (1976) Effects of angiotensin, vasopressin, and methoxamine on cardiac function and blood flow distribution in conscious dogs. Am J Physiol 231:1579 – 1587

Heyndrickx GR, Muylaert P, Pannier JL (1982) α-adrenergic control of oxygen delivery to myocardium during exercise in conscious dogs. Am J Physiol 242:H805 – H809

Heyndrickx GR, Vilaine JP, Moerman EJ, Leusen I (1984) Role of prejunctional alpha 2-adrenergic receptors in the regulation of myocardial performance during exercise in conscious dogs. Circ Res 54:683 – 693

Hibbs JB, Vavrin Z, Taintor RR (1987) L-arginine is required for expression of the activated macrophage effector mechanism causing selective metabolic inhibition in target cells. J Immunol 138:550 – 565

Hilton R, Eichholtz F (1925) The influence of chemical factors on the coronary circulation. J Physiol 59:413 – 425

Hirsch EF, Borghard-Erdle AM (1961) The innervation of the human heart. Arch Pathol 71:384 – 407

Hirsh PD, Hillis LD, Campbell WB, Firth BG, Willerson JT (1981) Release of prostaglandins and thromboxane into the coronary circulation in patients with ischemic heart disease. N Engl J Med 304:685 – 691

Hodgson JMcB, Marshall JJ (1989) Direct vasoconstriction and endothelium-dependent vasodilation. Mechanisms of acetylcholine effects on coronary flow and arterial diameter in patients with nonstenotic coronary arteries. Circulation 79:1043 – 1051

Hodgson JMcB, Cohen MD, Szentpetery S, Thames MD (1989) Effects of regional α- and β-blockade on resting and hyperemic coronary blood flow in conscious, unstressed humans. Circulation 79:797 – 809

Hoffman BB, Lefkowitz RJ (1980) Alpha-adrenergic receptor subtypes. N Engl J Med 302:1390 – 1396

Hoffman JIE (1987) Transmural myocardial perfusion. Prog Cardiovasc Dis 29:429 – 464

Holmgren S, Abrahamsson T, Almgren O (1985) Adrenergic innervation of coronary arteries and ventricular myocardium in the pig: fluorescence microscopic appearance in the normal state and after ischemia. Basic Res Cardiol 80:18 – 26

Holtz J, Grunewald WA, Manz R, von Restorff W, Bassenge E (1977a) Intracapillary hemoglobin oxygen saturation and oxygen consumption in different layers of the left ventricular myocardium. Pfluegers Arch 370:253 – 258

Holtz J, Mayer E, Bassenge E (1977b) Demonstration of alpha-adrenergic coronary control in different layers of canine myocardium by regional myocardial sympathectomy. Pfluegers Arch 372:187 – 194

Holtz J, Saeed M, Sommer O, Bassenge E (1982) Norepinephrine constricts the canine coronary bed via postsynaptic α_2-adrenoceptors. Eur J Pharmacol 82:199 – 202

Holtz J, Giesler M, Bassenge E (1983) Two dilatory mechanisms of anti-anginal drugs on epicardial coronary arteries in vivo: indirect, flow-dependent, endothelium-mediated dilation and direct smooth muscle relaxation. Z Kardiol 72 (Suppl 3):98–106

Holtz J, Förstermann U, Pohl U, Giesler M, Bassenge E (1984) Flow-dependent, endothelium-mediated dilation of epicardial coronary arteries in conscious dogs: effects on cyclooxygenase inhibition. J Cardiovasc Pharmacol 6:1161–1169

Holtz J, Busse R, Sommer O, Bassenge E (1987) Dilation of epicardial arteries in conscious dogs induced by angiotensin-converting enzyme inhibition with enalaprilat. J Cardiovasc Pharmacol 9:348–355

Homcy CJ, Graham RM (1985) Molecular characterization of adrenergic receptors. Circ Res 56:635–650

Hopwood AM, Burnstock G (1987) ATP mediates coronary vasoconstriction via P2x-purinoceptors and coronary vasodilatation via P2y-purinoceptors in the isolated perfused rat heart. Eur J Pharmacol 136:49–54

Horio Y, Yasue H, Rokutanda M, Makamura N, Ogawa H, Takaoka K, Matsuyama K, Kimura T (1986) Effects of intracoronary injection of acetylcholine on coronary arterial diameter. Am J Cardiol 57:984–989

Hossack KF, Brown BG, Stewart DK, Dodge HT (1984) Diltiazem-induced blockade of sympathetically mediated constriction of normal and diseased coronary arteries: lack of epicardial coronary dilatory effect in humans. Circulation 70:465–471

Houston DS, Shepherd JT, Vanhoutte PM (1986) Aggregating human platelets cause direct contraction and endothelium-dependent relaxation of isolated canine coronary arteries. Role of serotonin, thromboxane A2, and adenine nucleotides. J Clin Invest 78:539–544

Howes LG, Krum H (1989) Plasma lipoproteins, cardiovascular reactivity and the sympathetic nervous system. J Auton Pharmacol 9:293–301

Huang AH, Feigl EO (1988) Adrenergic coronary vasoconstriction helps maintain uniform transmural blood flow distribution during exercise. Circ Res 62:286–298

Igarashi Y, Aizawa Y, Tamura M, Ebe K, Yamaguchi T, Shibata A (1989) Vasoconstrictor effect of endothelin on the canine coronary artery: is a novel endogenous peptide involved in regulating myocardial blood flow and coronary spasm? Am Heart J 118:674–678

Ignarro LJ (1989) Biological actions and properties of endothelium-derived nitric oxide formed and released from artery and vein. Circ Res 65:1–21

Ignarro LJ, Byrns RE, Buga GM, Wood KS (1987) Endothelium-derived relaxing factor from pulmonary artery and vein possesses pharmacologic and chemical properties identical to those of nitric oxide radical. Circ Res 61:866–879

Inoue T, Tomoike H, Hisano K, Nakamura M (1988) Endothelium determines flow-dependent dilation of the epicardial coronary artery in dogs. J Am Coll Cardiol 11:187–191

Ito BR, Feigl EO (1985a) Carotid baroreceptor reflex coronary vasodilation in the dog. Circ Res 56:486–495

Ito BR, Feigl EO (1985b) Carotid chemoreceptor reflex parasympathetic coronary vasodilation in the dog. Am J Physiol 249:H1167–H1175

Jacobs M, Plane F, Bruckdorfer KR (1990) Native and oxidized low-density lipoproteins have different inhibitory effects on endothelium-derived relaxing factor in the rabbit aorta. Br J Pharmacol 100:21–26

Jie K, van Brummelen P, Vermey P, Timmermans PBMWM, van Zwieten PA (1984) Identification of vascular postsynaptic α_1- and α_2-adrenoceptors in man. Circ Res 54:447–452

Jie K, van Brummelen P, Vermey P, Timmermans PBMWM, van Zwieten PA (1987) Postsynaptic alpha 1- and alpha 2-adrenoceptors in human blood vessels: interactions with exogenous and endogenous catecholamines. Eur J Clin Invest 17:174–181

Johannsen UJ, Mark AL, Marcus ML (1982) Responsiveness to cardiac sympathetic nerve stimulation during maximal coronary dilation produced by adenosine. Circ Res 50:510–517

Johansson B, Mellander S (1975) Static and dynamic components in the vascular myogenic response to passive changes in length as revealed by electrical and mechanical recordings from the rat portal vein. Circ Res 36:76–83

Johnson JR, Di Palma JR (1939) Intramyocardial pressure and its relation to aortic blood pressure. Am J Physiol 125:234–243

Jones CE, Liang IYS, Maulsby MR (1986) Cardiac and coronary effects of prazosin and phenoxybenzamine during coronary hypotension. J Pharmacol Exp Ther 236:204–211

Jugdutt BI (1981) Effects of prostacyclin, prostaglandin E1 and E2 on spontaneous ventricular arrhythmias in infarction. Circulation 64 (Suppl IV):IV–319 (abstract)

Jugdutt BI, Hutchins GM, Bulkley BH, Becker LC (1980) Salvage of ischemic myocardium by ibuprofen during infarction in the conscious dog. Am J Cardiol 46:74–82

Jugdutt BI, Hutchins GM, Bulkley BH, Becker LC (1981) Dissimilar effects of prostacyclin, prostaglandin E1 and prostaglandin E2 on myocardial infarct size after coronary occlusion in conscious dogs. Circ Res 49:685–700

Kalsner S (1985) Cholinergic mechanisms in human coronary artery preparations: implications of species differences. J Physiol 358:509–526

Kamiya A, Togawa T (1980) Adaptive regulation of wall shear stress to flow change in the canine carotid artery. Am J Physiol 239:H14–H21

Kanatsuka H, Lamping KG, Eastham CL, Marcus ML (1990) Heterogenous changes in epimyocardial microvascular size during graded coronary stenosis. Evidence of the microvascular site for autoregulation. Circ Res 66:389–396

Kasuya Y, Takuwa Y, Yanagisawa M, Kimura S, Goto K, Masaki T (1989) Endothelin-1 induces vasoconstriction through two functionally distinct pathways in porcine coronary artery: contribution of phosphoinositide turnover. Biochem Biophys Res Comm 161:1049–1055

Katori M, Berne RM (1966) Release of adenosine from anoxic hearts. Relationship to coronary flow. Circ Res 19:420–425

Kelley KO, Feigl EO (1978) Segmental alpha-receptor-mediated vasoconstriction in the canine coronary circulation. Circ Res 43:908–917

Kelm M, Schrader J (1988) Nitric oxide release from the isolated guinea pig heart. Eur J Pharmacol 155:317–321

Khayyal MA, Eng C, Franzen D, Breall JA, Kirk ES (1985) Effects of vasopressin on the coronary circulation: reserve and regulation during ischemia. Am J Physiol 248: H516–H522

Kjekshus JK (1973) Mechanisms for flow distribution in normal and ischemic myocardium during increased ventricular preload in the dog. Circ Res 33:489–499

Klocke FJ (1987) Measurements of coronary flow reserve: defining pathophysiology versus making decisions about patient care. Circulation 76:1183–1189

Klocke FJ, Kaiser GA, Ross Jr J, Braunwald E (1965) An intrinsic adrenergic vasodilator mechanism in the coronary vascular bed of the dog. Circ Res 16:376–382

Klocke FJ, Weinstein IR, Klocke JF, Ellis AK, Kraus DR, Mates RE, Canty JM, Anbar RD, Romanowski RR, Wallmeyer KW, Echt MP (1981) Zero-flow pressures and pressure-flow relationships during single long diastoles in the canine coronary bed before and during maximum vasodilation. J Clin Invest 68:970–980

Knabb RM, Ely SW, Bacchus AN, Rubio R, Berne RM (1983) Consistent parallel relationships among myocardial oxygen consumption, coronary blood flow, and pericardial infusate adenosine concentration with various interventions and β-blockade in the dog. Circ Res 53:33–41

Knowles RG, Palacios M, Palmer RMJ, Moncada S (1989) Formation of nitric oxide from L-arginine in the central nervous system: a transduction mechanism for stimulation of the soluble guanylate cyclase. Proc Natl Acad Sci USA 86:5159–5162

Kobinger W, Pichler L (1980) Investigation into different types of post- and presynaptic α-adrenoceptors at cardiovascular sites in rats. Eur J Pharmacol 65:393–402

Kobinger W, Pichler L (1981) α_1- and α_2-adrenoceptor subtypes: selectivity of various agonists and relative distribution of receptors as determined in rats. Eur J Pharmacol 73:313–321

Kodama M, Kanaide H, Abe S, Hirano K, Kai H, Nakamura M (1989) Endothelin-induced Ca-independent contraction of the porcine coronary artery. Biochem Biophys Res Comm 160:1302–1308

Koltai MZ, Rösen P, Hadhazy P, Ballagi-Pordany G, Köszeghy A, Pogatsa G (1988) Relationship between vascular adrenergic receptors and prostaglandin biosynthesis in canine diabetic coronary arteries. Diabetologia 31: 681–686

Kopia GA, Kopaciewicz LJ, Ruffolo Jr RR (1986) Alpha adrenoceptor regulation of coronary artery blood flow in normal and stenotic canine coronary arteries. J Pharmacol Exp Ther 239:641–647

Koster PF, Ohlstein EH, Nichols AJ (1989) The effect of intracoronary endothelin on coronary hemodynamics and cardiac function in anesthetized dogs. Faseb J 3:A878 (abstract)

Kreuzer H, Schoeppe W (1963) Das Verhalten des Druckes in der Herzwand. Pfluegers Arch 278:181–198

Kroll K, Feigl EO (1985) Adenosine is unimportant in controlling coronary blood flow in unstressed dog hearts. Am J Physiol 249:H1176–H1187

Kroll K, Schrader J, Piper HM, Henrich M (1987) Release of adenosine and cyclic AMP from coronary endothelium in isolated guinea pig hearts: relation to coronary flow. Circ Res 60:659–665

Ku DD (1989) Endothelin produces potent vasoconstriction in isolated human coronary arteries, veins and long-term coronary aortic bypass grafts. Circulation 80 (Suppl II):II–213 (abstract)

Kulakowski EC, Lampson WG, Schaffer SW, Lovenberg W (1983) Action of substance P on the working rat heart. Biochem Pharmacol 32:1097–1100

Kumpuris AG, Luchi RJ, Waddell CC, Miller RR (1980) Production of circulating platelet aggregates by exercise in coronary patients. Circulation 61:62–65

Kuo L, Chilian WM, Davis MJ (1990) Coronary arteriolar myogenic response is independent of endothelium. Circ Res 66:860–866

Kurihara H, Yamaoki K, Nagai R, Yoshizumi M, Takaku F, Satoh H, Inui J, Yazaki Y (1989) Endothelin – a potent vasoconstrictor associated with coronary vasospasm. Life Sci 44:1937–1943

Kuzuya T, Tada M, Ohmori M, Inui M, Abe H, Yamagishi M, Kodama K (1981) Altered metabolism of thromboxane A2 and prostaglandin I2 in patients with angina pectoris. Circulation 64 (Suppl IV):IV–143 (abstract)

Lam LYT, Chesebro JH, Steele PM, Badimon L, Fuster V (1987) Is vasospasm related to platelet deposition? Relationship in a porcine preparation of arterial injury in vivo. Circulation 75:243–248

Lamping KG, Dole WP (1987) Acute hypertension selectively potentiates constrictor responses of large coronary arteries to serotonin by altering endothelial function in vivo. Circ Res 61:904–913

Lamping KG, Eastham CL (1989) Endothelin: a potent vasoconstrictor in the coronary microcirculation. Circulation 80 (Suppl II):II–212 (abstract)

Lamping KG, Kanatsuka H, Eastham CL, Chilian WM, Marcus ML (1989) Nonuniform vasomotor responses of the coronary microcirculation to serotonin and vasopressin. Circ Res 65:343–351

Langer GA, Brady AJ (1966) Potassium in dog ventricular muscle: kinetic studies of distribution and effects of varying frequency of contraction and potassium concentration of perfusate. Circ Res 18:164–177

Langer SZ (1981) Presynaptic regulation of the release of catecholamines. Pharmacol Rev 32:337–362

Langer SZ, Adler-Graschinsky E, Giorgi O (1977) Physiological significance of α-adreno-ceptor-mediated negative feedback mechanism regulating noradrenaline release during nerve stimulation. Nature 265:648–650

Langille BL, O'Donnel F (1986) Reductions in arterial diameter produced by chronic decreases in blood flow are endothelium-dependent. Science 231:405–407

Lansman JB, Hallam TJ, Rink TJ (1987) Single stretch-activated ion channels in vascular endothelial cells as mechanotransducers? Nature 325:811–813

Larkin SW, Clarke JG, Koegh BE, Araujo L, Rhodes C, Davies GJ, Taylor KM, Maseri A (1989) Intracoronary endothelin induces myocardial ischemia by small vessel constriction in the dog. Am J Cardiol 64:956–958

Laxson DD, Dai X-Z, Homans DC, Bache RJ (1989) The role of α_1- and α_2-adrenergic receptors in mediation of coronary vasoconstriction in hypoperfused ischemic myocardium during exercise. Circ Res 65:1688–1697

Levene DL, Freeman MR (1976) α-adrenoceptor-mediated coronary artery spasm. J Am Med Assoc 236:1018–1022

Lewis HD, Davis JW, Archibald DG, Steinke WE, Smitherman TC, Doherty JE, Schnaper HW, Le Winter MM, Linares E, Pouget JM, Sabharwal SC, Chesler E, De Mots H (1983) Protective effects of aspirin against acute myocardial infarction and death in men with unstable angina. Results of a veterans administration cooperative study. N Engl J Med 309:396–403

Liang C-S, Gavras H, Black J, Sherman LG, Hood Jr WB (1982) Renin-angiotensin system inhibition in acute myocardial infarction in dogs. Circulation 66:1249–1255

Liang IYS, Jones CE (1985) Alpha 1-adrenergic blockade increases coronary blood flow during coronary hypoperfusion. Am J Physiol 249:H1070–H1077

Linder C, Heusch G (1990) ACE-inhibitors for the treatment of myocardial ischemia? Cardiovasc Drugs Ther (in press)

Linder L, Kiowski W, Bühler FR, Lüscher TF (1990) Indirect evidence for the release of endothelium-derived relaxing factor in the human forearm circulation in vivo: blunted response in essential hypertension. Circulation 81:1762–1767

Liu JJ, Chen DJ, Casley DJ, Nayler WG (1990a) Effect of ischaemia and reperfusion on ^{125}I endothelin binding in rat cardiac membranes. Am J Physiol 258:H829–H835

Lochner W, Nasseri M (1959) Über den venösen Sauerstoffdruck, die Einstellung der Coronardurchblutung und den Kohlenhydratstoffwechsel des Herzens bei Muskelarbeit. Pfluegers Arch 269:407–416

Lopez JAG, Armstrong ML, Piegors DJ, Heistad DD (1989) Effect of early and advanced atherosclerosis on vascular responses to serotonin, thromboxane A2, and ADP. Circulation 79:698–705

Ludmer PL, Selwyn AP, Shook TL, Wayne RR, Mudge GH, Alexander RW, Ganz P (1986) Paradoxial vasoconstriction induced by acetylcholine in atherosclerotic coronary arteries. N Engl J Med 315:1046–1051

Lückhoff A, Busse R (1990) Calcium influx into endothelial cells and formation of EDRF is controlled by the membrane potential. Pfluegers Arch 416:305–311

Lüscher TF, Vanhoutte PM (1986) Endothelium-dependent contractions to acetylcholine in the aorta of the spontaneously hypertensive rat. Hypertension 8:344–348

Lüscher TF, Rau L, Vanhoutte PM (1987) Endothelium-dependent vascular responses in normotensive and hypertensive Dahl rats. Hypertension 9:157–163

Marcus ML, Wright C, Doty D, Eastham C, Laughlin D, Krumm P, Fastenow C, Brody M (1981) Measurements of coronary velocity and reactive hyperemia in the coronary circulation of humans. Circ Res 49:877–891

Margolius HS (1989) Tissue kallikreins and kinins: regulation and roles in hypertensive and diabetic diseases. Annu Rev Pharmacol Toxicol 29:343–364

Mark AL, Abboud FM, Schmid PG, Heistad DD, Mayer UJ (1972) Differences in direct effects of adrenergic stimuli on coronary, cutaneous and muscular vessels. J Clin Invest 51:279–287

Marletta MA, Yoon PS, Iyengar R, Leaf CD, Wishnok JS (1988) Macrophage oxidation of L-arginine to nitrite and nitrate: nitric oxide is an intermediate. Biochemistry 27:8706–8711

Marsden PA, Danthuluri NR, Brenner BM, Ballermann BJ, Brock TA (1989) Endothelin action on vascular smooth muscle involves inositol triphosphate and calcium mobilization. Biochem Biophys Res Comm 158:86–93

Martin SE, Patterson RE (1989) Coronary constriction due to neuropeptide Y: alleviation with cyclooxygenase blockers. Am J Physiol 257:H927–H934

Maruoka Y, McKirnan MD, Engler RL, Longhurst JC (1987) Functional significance of alpha-adrenergic receptors in mature coronary collateral circulation of dogs. Am J Physiol 253:H582–H590

Mates RE, Klocke FJ, Canty JM (1988) Coronary capacitance. Prog Cardiovasc Dis 31:1–15

Matsuda H, Kuon E, Holtz J, Busse R (1985) Endothelium-mediated dilations contribute to the polarity of the arterial wall in vasomotion induced by alpha 2-adrenergic agonists. J Cardiovasc Pharmacol 7:680–688

Matsuzaki M, Patritti J, Tajimi T, Miller M, Kemper WS, Ross Jr J (1984) Effects of β-blockade on regional myocardial flow and function during exercise. Am J Physiol 247:H52–H60

McCall TB, Boughton-Smith NK, Palmer RMJ, Whittle BJR, Moncada S (1989) Synthesis of nitric oxide from L-arginine by neutrophils. Biochem J 261:293–296

McEwan J, Larkin S, Davies G, Chierchia S, Brown M, Stevenson J, MacIntyre I, Maseri A (1986) Calcitonin gene-related peptide: a potent dilator of human epicardial coronary arteries. Circulation 74:1243–1247

McGrath JC (1982) Evidence for more than one type of postjunctional alpha-adrenoceptor. Biochem Pharmacol 31:467–484

McHale PA, Dube GP, Greenfield Jr JC (1987) Evidence for myogenic vasomotor activity in the coronary circulation. Prog Cardiovasc Dis 30:139–146

McRaven DR, Mark AL, Abboud FM, Mayer HE (1971) Responses of coronary vessels to adrenergic stimuli. J Clin Invest 50:773–778

Melkumyants AM, Balashov SA, Veselova ES, Khayutin VM (1987) Continuous control of the lumen of feline conduit arteries by blood flow rate. Cardiovasc Res 21:863–870

Melkumyants AM, Balashov SA, Khayutin VM (1989) Endothelium dependent control of arterial diameter by blood viscosity. Cardiovasc Res 23:741–747

Mellander S, Johansson B (1968) Control of resistance, exchange, and capacitance functions in the peripheral circulation. Pharmacol Rev 20:117–196

Mellander S, Johansson B, Gray S, Jonsson O, Lundvall J, Ljung B (1967) The effects of hyperosmolarity on intact and isolated vascular smooth muscle. Possible role in exercise hyperemia. Angiologica 4:310–322

Miller WL, Belardinelli L, Bacchus A, Foley DH, Rubio R, Berne RM (1979) Canine myocardial adenosine and lactate production, oxygen consumption, and coronary blood flow during stellate ganglia stimulation. Circ Res 45:708–718

Miwa K, Kambara H, Kawai C (1981) Exercise-induced angina evoked by aspirin administration in patients with variant angina. Am J Cardiol 47:1210–1214

Miyauchi T, Yanagisawa M, Tomizawa T, Sugishita Y, Suzuki N, Fujino M, Ajisaka R, Goto K, Masaki T (1989) Increased plasma concentrations of endothelin-1 and big endothelin-1 in acute myocardial infarction. Lancet II:53–54

Mohrman DE, Feigl EO (1978) Competition between sympathetic vasoconstriction and metabolic vasodilation in the canine coronary circulation. Circ Res 42:79–86

Moncada S, Gryglewski R, Bunting S, Vane JR (1976) An enzyme isolated from arteries transforms prostaglandin endoperoxides to an unstable substance that inhibits platelet aggregation. Nature 263:663–665

Moncada S, Palmer RMJ, Higgs EA (1989) Biosynthesis of nitric oxide from L-arginine. A pathway for the regulation of cell function and communication. Biochem Pharmacol 38:1709–1715

Monos E, Cox RH, Peterson LH (1978) Direct effect of physiological doses of arginine vasopressin on the arterial wall in vivo. Am J Physiol 234:H167–H172

Mosher P, Ross Jr J, McFate PA, Shaw RF (1964) Control of coronary blood flow by an autoregulatory mechanism. Circ Res 14:250–259

Motulsky HJ, Snavely MD, Hughes RJ, Insel PA (1983) Interaction of verapamil and other calcium channel blockers with α_1- and α_2-adrenergic receptors. Circ Res 52:226–231

Motulsky HJ, Maisel AS, Snavely MD, Insel PA (1984) Quinidine is a competitive antagonist at alpha 1- and alpha 2-adrenergic receptors. Circ Res 55:376–381

Mudge GH, Grossman W, Mills Jr RM, Lesch M, Braunwald E (1976) Reflex increase in coronary vascular resistance in patients with ischemic heart disease. N Engl J Med 295:1333–1337

Mudge GH, Goldberg S, Gunther S, Mann T, Grossman W (1979) Comparison of metabolic and vasoconstrictor stimuli on coronary vascular resistance in man. Circulation 59:544–550

Mueller HS, Rao PS, Rao PB, Gory DJ, Mudd JG, Ayres SM (1982) Enhanced transcardiac l-norepinephrine response during cold pressor test in obstructive coronary artery disease. Am J Cardiol 50:1223–1228

Mülsch A, Bassenge E, Busse R (1989) Nitric oxide synthesis in endothelial cytosol: evidence for a calcium-dependent and a calcium-independent mechanism. Naunyn Schmiedebergs Arch Pharmacol 340:767–770

Mülsch A, Busse R (1990) NG-nitro-L-arginine (N5-[imino (nitroamino) methyl]-L-ornithine) impairs endothelium-dependent dilations by inhibiting cytosolic nitric oxide synthesis from L-arginine. Naunyn Schmiedebergs Arch Pharmacol 341:143–147

Münzel T, Stewart DJ, Holtz J, Bassenge E (1988) Preferential venoconstriction by cyclooxygenase inhibition in vivo without attenuation of nitroglycerin venodilation. Circulation 78:407–415

Murphree SS, Saffitz JE (1988) Delineation of the distribution of β-adrenergic receptor subtypes in canine myocardium. Circ Res 63:117–125

Murray PA, Vatner SF (1979) α-adrenoceptor attenuation of coronary vascular response to severe exercise in the conscious dog. Circ Res 45:654–660

Murray PA, Belloni FL, Sparks HV (1979) The role of potassium in the metabolic control of coronary vascular resistance of the dog. Circ Res 44:767–780

Murray PA, Lavallee M, Vatner SF (1984) Alpha-adrenergic-mediated reduction in coronary blood flow secondary to carotid chemoreceptor reflex activation in conscious dogs. Circ Res 54:96–106

Mustard JF, Kinlough-Rathbone RL, Packham MA (1983) Aspirin in the treatment of cardiovascular disease: a review. Am J Med 74 (Suppl A6):43–49

Myers PR, Banitt PF, Guerra R, Harrison DG (1989) Characteristics of canine coronary resistance arteries: importance of endothelium. Am J Physiol 257:H603–H610

Myers P, Minor R, Guarra R, Bates J, Harrison D (1990) Vasorelaxant properties of the endothelium-derived relaxing factor more closely resemble S-nitrosocysteine than nitric oxide. Nature 345: (in press)

Nabel EG, Ganz P, Gordon JB, Alexander RW, Selwyn AP (1988a) Dilation of normal and constriction of atherosclerotic coronary arteries caused by the cold pressor test. Circulation 77:43–52

Nabel EG, Ganz P, Selwyn AP (1988b) Atherosclerosis impairs flow-mediated dilation in human coronary arteries. Circulation 78 (Suppl II):II–474 (abstract)

Nagasawa K, Tomoike H, Hayashi Y, Yamada A, Yamamoto T, Nakamura M (1989) Intramural hemorrhage and endothelial changes in atherosclerotic coronary artery after repetitive episodes of spasm in x-ray-irradiated hypercholesterolemic pigs. Circ Res 65:272–282

Nagata M, Pichet R, Lavallee M (1988) Coronary dilation with carotid chemoreceptor stimulation in cardiac-denervated dogs. Am J Physiol 255:H1330–1335

Nakane T, Chiba S (1987) Postjunctional α-adrenoceptor subtypes in isolated and perfused canine epicardial coronary arteries. J Cardiovasc Pharmacol 10:651–657

Nakane T, Tsujimoto G, Hashimoto K, Chiba S (1988) Beta adrenoceptors in the canine large coronary arteries: beta-1 adrenoceptors predominate in vasodilation. J Pharmacol Exp Ther 245:936–943

Nakao K, Saito Y, Matsuyama K, Okumura K, Jougasaki M, Yasue H (1989) Implication of endothelin in variant angina. Circulation 80 (Suppl II):II–586 (abstract)

Nathan HJ, Feigl EO (1986) Adrenergic vasoconstriction lessens transmural steal during coronary hypoperfusion. Am J Physiol 250:H645–H653

Nellessen U, Lee TC, Fischell TA, Ginsburg R, Masuyama T, Alderman EL, Schroeder JS (1988) Effects of acetylcholine on epicardial coronary arteries after cardiac transplantation without angiographic evidence of fixed graft narrowing. Am J Cardiol 62:1093–1097

Nichols WW, Mehta JL, Thompson L, Donnelly WH (1988) Synergistic effects of LTC4 and TxA2 on coronary flow and myocardial function. Am J Physiol 255:H153–H159

Nicod P, Winniford MD, Campbell WB, Rehr RB, Firth BG, Hillis LD (1984) Alterations in coronary blood flow induced by cigarette smoking: Lack of relation to plasma arginine vasopressin concentrations. Am J Cardiol 54:667–668

Olesen S-P, Clapham DE, Davies PF (1988) Haemodynamic shear stress activates a K+ current in vascular endothelial cells. Nature 331:168–170

Olsson RA, Bünger R (1987) Metabolic control of coronary blood flow. Prod Cardiovasc Dis 29:369–387

Orlick AE, Ricci DR, Alderman EL, Stinson EB, Harrison DC (1978) Effects of alpha adrenergic blockade upon coronary hemodynamics. J Clin Invest 62:459–467

Osborne JA, Siegman MJ, Sedar AW, Mooers SU, Lefer AM (1989a) Lack of endothelium-dependent relaxation in coronary resistance arteries of cholesterol-fed rabbits. Am J Physiol 256:C591–C597

Osborne JA, Lento PH, Siegfried MR, Stahl GL, Fusman B, Lefer AM (1989b) Cardiovascular effects of acute hypercholesterolemia in in rabbits. J Clin Invest 83:465–473

Palmer RMJ, Ferrige AG, Moncada S (1987) Nitric oxide release accounts for the biological activity of endothelium-derived relaxing factor. Nature 327:524–526

Pantely GA, Bristow JD, Swenson LJ, Ladley HD, Johnson WB, Anselone CG (1985a) Incomplete coronary vasodilation during myocardial ischemia in swine. Am J Physiol 249:H638–H647

Pantely GA, Ladley HD, Anselone CG, Bristow JD (1985b) Vasopressin-induced coronary constriction at low perfusion pressures. Cardiovasc Res 19:433–441

Panzenbeck MJ, Kaley G (1983) Leukotriene D4 reduces coronary blood flow in the anesthetized dog. Prostaglandins 25:661–670

Peters KG, Marcus ML, Harrison DG (1989) Vasopressin and the mature coronary collateral circulation. Circulation 79:1324–1331

Pfeffer MA, Pfeffer JM, Steinberg C, Finn P (1985) Survival after an experimental myocardial infarction: beneficial effects of long-term therapy with captopril. Circulation 72:406–412

Pitt B, Sugishita Y, Gregg DE (1969a) Coronary hemodynamic effects of calcium in the unanesthetized dog. Am J Physiol 216:1456–1459

Pitt B, Mason J, Conti CR, Colman RW (1969b) Activation of the plasma kallikrein during myocardial ischemia. Pharmacol Res Comm 1:185–186

Pitt B, Pasyk S, Walton J, Grekin R (1982) Endogenous arginine vasopressin release in patients with coronary artery spasm. Circulation 66 (Suppl II):II–88 (abstract)

Pitt B, Shea MJ, Romson JL, Lucchesi BR (1983) Prostaglandins and prostaglandin inhibitors in ischemic heart disease. Ann Intern Med 99:83–92

Pohl U, Busse R (1988) Reduced nutritional blood flow in autoperfused rabbit hindlimbs following inhibition of endothelial vasomotor function. In: Halpern W, Pegram B, Brayden J, Mackey K, McLaughlin M, Osol G (eds) Resistance arteries. Perinatology Press, Ithaca, NY, pp 10–16

Pohl U, Busse R (1989a) EDRF increases cyclic GMP in platelets during passage through the coronary vascular bed. Circ Res 65:1798–1803

Pohl U, Busse R (1989b) Hypoxia stimulates release of endothelium-derived relaxant factor. Am J Physiol 256:H1595–H1600

Pohl U, Busse R (1989c) Differential vascular sensitivity to luminally and adventitially applied endothelin-1. J Cardiovasc Pharmacol 13 [Suppl 5]:S188–S190

Pohl U, Holtz J, Busse R, Bassenge E (1986a) Crucial role of endothelium in the vasodilator response to increased flow in vivo. Hypertension 8:37–44

Pohl U, Busse R, Kuon E, Bassenge E (1986b) Pulsatile perfusion stimulates the release of endothelial autacoids. J Appl Cardiol 1:215–235

Pohl U, Herlan K, Huang A, Bassenge E (1990a) EDRF-mediated, shear-induced dilation opposes myogenic vasoconstriction. Am J Physiol (in press)

Pohl U, Lamontagne D, Bassenge E, Busse R (1990b) EDRF augments coronary conductivity through attenuation of myogenic autoregulation. Pflügers Arch 414 Suppl 1:R62 (abstract)

Quadt JFA, Voss R, TenHoor F (1982) Prostacyclin production of the isolated pulsatingly perfused rat aorta. J Pharmacol Method 7:263–270

Raberger G, Weissel M, Kraupp O (1971) The dependence of the effects of i. cor. administered adenosine and of coronary conductance on the arterial pH, pCO_2 and buffer capacity in dogs. Naunyn Schmiedebergs Arch Pharmacol 271:301–310

Radomski MW, Palmer RMJ, Moncada S (1987a) The role of nitric oxide and cGMP in platelet adhesion to vascular endothelium. Biochem Biophys Res Comm 148:1482–1489

Radomski MW, Palmer RMJ, Moncada S (1987b) Comparative pharmacology of endothelium-derived relaxing factor, nitric oxide and prostacyclin in platelets. Br J Pharmacol 92:181–187

Radomski MW, Palmer RMJ, Moncada S (1987c) The anti-aggregating properties of vascular endothelium: interactions between prostacyclin and nitric oxide. Br J Pharmacol 92:639–646

Raff WK, Lochner W (1974) Wirkungsmechanismus von Nitroglycerin. Med Klin 69:1100–1104

Raff WK, Kosche F, Lochner W (1971a) Herzfrequenz und extravasale Komponente des Coronarwiderstandes. Pfluegers Arch 323:241–249

Raff WK, Kosche F, Lochner W (1971b) Extravasale Komponente des Coronarwiderstandes und Coronardurchblutung bei steigendem enddiastolischen Druck. Pfluegers Arch 327:225–233

Raff WK, Kosche F, Lochner W (1971c) Die extravasale Komponente des Coronarwiderstandes bei Steigerung der linksventrikulären Druckanstiegsgeschwindigkeit durch Isoproterenol. Pfluegers Arch 325:323–333

Raff WK, Kosche F, Lochner W (1972a) Extravascular coronary resistance and its relation to microcirculation. Am J Cardiol 29:598–603

Raff WK, Kosche F, Goebel H, Lochner W (1972b) Coronary extravascular resistance at increasing left ventricular pressure. Pfluegers Arch 333:352–361

Rafflenbeul W, Bassenge E, Lichtlen PR (1988) Competition between endothelium- and nitroglycerin-induced coronary vasodilation. Circulation 78 (Suppl II):II–455 (abstract)

Raizner AE, Chahine RA, Ishimori T, Verani MS, Zacca N, Jamal N, Miller RR, Luchi RJ (1980) Provocation of coronary artery spasm by the cold pressor test. Circulation 62:925–932

Rees DD, Palmer RM, Moncada S (1989) Role of endothelium-derived nitric oxide in the regulation of blood pressure. Proc Natl Acad Sci USA 86:3375–3378

Reid JVO, Ito BR, Huang AH, Buffington CW, Feigl EO (1985) Parasympathetic control of transmural coronary blood flow in dogs. Am J Physiol 249:H337–H343

Richardt G, Waas W, Kranzhöfer R, Mayer E, Schömig A (1987) Adenosine inhibits exocytotic release of endogeneous noradrenaline in the rat heart: A protective mechanism in early myocardial ischemia. Circ Res 61:117–123

Richardt G, Waas W, Kranzhöfer R, Cheng B, Lohse MJ, Schömig A (1989) Interaction between the release of adenosine and noradrenaline during sympathetic stimulation: A feedback mechanism in rat heart. J Mol Cell Cardiol 21:269–277

Rimele TJ, Rooke TW, Aarhus LL, Vanhoutte PM (1983) Alpha-1 adrenoceptors and calcium in isolated canine coronary arteries. J Pharmacol Exp Ther 226:668–672

Rinkema LE, Thomas Jr JX, Randall WC (1982) Regional coronary vasoconstriction in response to stimulation of stellate ganglia. Am J Physiol 243:H410–H415

Robertson RM, Robertson D, Roberts LJ, Maas RL, FitzGerald GA, Friesinger GC, Oates JA (1981) Thromboxane A2 in vasotonic angina pectoris. Evidence from direct measurements and inhibitor trials. N Engl J Med 304:998–1003

Robertson RM, Bernard YD, Carr RK, Robertson D (1983) Alpha-adrenergic blockade in vasotonic angina: lack of efficacy of specific alpha-receptor blockade with prazosin. J Am Coll Cardiol 2:1146–1150

Rodbard S (1975) Vascular caliber. Cardiology 60:4–49

Rosendorff C, Hoffman JIE, Verrier ED, Rouleau J, Boerboom LE (1981) Cholesterol potentiates the coronary artery response to norepinephrine in anesthetized and conscious dogs. Circ Res 48:320–329

Ross J, Klocke FJ, Kaiser G, Braunwald E (1963) Effect of alterations of coronary blood flow on the oxygen consumption of the working heart. Circ Res 13:510–513

Rouleau J, Boerboom LE, Surjadhana A, Hoffman JIE (1979) The role of autoregulation and tissue diastolic pressures in the transmural distribution of left ventricular blood flow in anesthetized dogs. Circ Res 45:804–815

Rousseau MF, Close P, Pouleur H (1989) Are the angiotensin-converting enzyme inhibitors poor anti-ischemic drugs? Circulation 80 (Suppl II):II–52 (abstract)

Rubanyi GM, Romero JC, Vanhoutte PM (1986) Flow-induced release of endothelium-derived relaxing factor. Am J Physiol 250:H1145–H1149

Rudehill A, Sollevi A, Franco-Cereceda A, Lundberg JM (1986) Neuropeptide Y (NPY) and the pig heart: release and coronary vasoconstrictor effects. Peptides 7:821–826

Ruffolo RR, Sulpizio AC, Nichols AJ, DeMarinis RM, Hieble JP (1987) Pharmacologic differentiation between pre- and postjunctional alpha 2-adrenoceptors by SKandF 104078. Naunyn Schmiedebergs Arch Pharmacol 336:415–418

Sabbah HN, Stein PD (1982) Effect of acute regional ischemia on pressure in the subepicardium and subendocardium. Am J Physiol 242:H240–H244

Sabiston DC, Gregg DE (1957) Effect of cardiac contraction on coronary blood flow. Circulation 15:14–20

Saeed M, Sommer O, Holtz J, Bassenge E (1982) α-adrenoceptor blockade by phentolamine causes β-adrenergic vasodilation by increased catecholamine release due to presynaptic α-blockade. J Cardiovasc Pharmacol 4:44–52

Saeed M, Holtz J, Elsner D, Bassenge E (1985) Sympathetic control of myocardial oxygen balance in dogs mediated by activation of coronary vascular α_2-adrenoceptors. J Cardiovasc Pharmacol 7:167–173

Saito D, Steinhart CR, Nixon DG, Olsson RA (1981) Intracoronary adenosine deaminase reduces canine myocardial reactive hyperemia. Circ Res 49:1262–1267

Sakuma I, Togashi H, Yoshioka M, Kobayashi T, Saito H, Yasuda H, Gross S, Levi R (1990) Effects of intracisternal and intracisternal administration of L-NG-monomethyl arginine on renal sympathetic nerve activity in anesthetized rats. In: Moncada S, Higgs EA (eds) Nitric oxide from L-arginine, a bioregulatory system. Elsevier, Amsterdam pp 481–482

Salminen K, Tikkanen I, Saijonmaa O, Nieminen M, Fyhrquist F, Frick MH (1989) Modulation of coronary tone in acute myocardial infarction by endothelin. Lancet II:747

Schipke J, Heusch G, Deussen A, Thaemer V (1985) Acetylcholine induces constriction of epicardial coronary arteries in anesthetized dogs after removal of endothelium. Drug Res 35:926–929

Schipke J, Heusch G, Thämer V (1987) Evidence against the adenosine catecholamine antagonism in the canine heart in situ. Drug Res 37:1345–1347

Schmid PG, Wendling MG, Mark AL, Eckstein JW, Abboud FM (1970) Coronary vascular response to physiological levels of vasopressin. Circulation 41/42 (Suppl III):III–196 (abstract)

Schmid PG, Abboud FM, Wendling MG, Ramberg ES, Mark AL, Heistad DD, Eckstein JW (1974) Regional vascular effects of vasopressin: plasma levels and circulatory responses. Am J Physiol 227:998–1004

Schmitz JM, Apprill PG, Buja M, Willerson JT, Campbell WB (1985) Vascular prostaglandin and thromboxane production in a canine model of myocardial ischemia. Circ Res 57:223–231

Schrader J (1981) Sites of action and production of adenosine in the heart. In: Burnstock G (ed) Purinergic receptors. Chapman and Hall, London, pp 121–162

Schrader J, Deussen A (1988) Free cytosolic adenosine sensitivity signals myocardial hypoxia. In: Acker H (ed) Oxygen sensing in tissue. Springer, Berlin Heidelberg New York, pp 165–176

Schrader J, Gerlach E (1976) Compartmentation of cardiac adenine nucleotides and formation of adenosine. Pfluegers Arch 367:129–135

Schrader J, Rubio R, Berne RM (1975) Inhibition of slow action potentials of guinea pig atrial muscle by adenosine: a possible effect on Ca- influx. J Mol Cell Cardiol 7:427–433

Schrader J, Baumann G, Gerlach E (1977) Antiadrenergic action of adenosine in the heart: possible physiological significance. Pfluegers Arch 372:29–35

Schrör K (1990) Thromboxane A2 and platelets as mediators of coronary arterial vasoconstriction in myocardial ischemia. Eur Heart J 11 (Suppl B):27–34

Schrör K, Ahland B, Weiss P, König E (1988) Stimulation of coronary vascular PGI_2 by organic nitrates. Eur Heart J 9 (Suppl A):25–32

Schulz R, Heusch G, Oudiz R, Guth BD (1989) Coronary pressure and flow have no independent effect on contractility in anesthetized swine. Faseb J 3:A405 (abstract)

Schulz R, Oudiz RJ, Guth BD, Heusch G (1990) Minimal α_1- and α_2-adrenoceptor mediated coronary vasconstriction in the anaesthetized swine. Naunyn Schmiedebergs Arch 342:422–428

Schumacher WA, Heran CL, Goldenberg HJ, Harris DN, Ogletree ML (1989) Magnitude of thromboxane receptor antagonism necessary for antithrombotic activity in monkeys. Am J Physiol 256:H726–H734

Schwartz GG, McHale PA, Greenfield JC (1982) Hyperemic response of the coronary circulation to brief diastolic occlusion in the conscious dog. Circ Res 50:28–37

Schwartz JS, Carlyle PF, Cohn JN (1980) Effect of coronary arterial pressure on coronary stenosis resistance. Circulation 61:70–76

Schwartz PJ, Stone HL (1977) Tonic influence of the sympathetic nervous system on myocardial reactive hyperemia and on coronary blood flow distribution in dogs. Circ Res 41:51–58

Scott JB, Radawski D (1971) Role of hyperosmolarity in the genesis of active and reactive hyperemia. Circ Res 28 (Suppl I):I26–I32

Seitelberger R, Schütz W, Schlappack O, Raberger G (1984) Evidence against the adenosine-catecholamine antagonism under in vivo conditions. Naunyn Schmiedebergs Arch 325:234–239

Seitelberger R, Guth BD, Lee JD, Katayama K, Heusch G, Ross Jr J (1986) Alpha 1 and alpha 2 receptor stimulation in conscious dogs increase coronary resistance but not myocardial function. J Am Coll Cardiol 7 (Suppl A):81A (abstract)

Seitelberger R, Guth BD, Heusch G, Lee JD, Katayama K, Ross Jr J (1988) Intracoronary alpha 2-adrenergic receptor blockade attenuates ischemia in conscious dogs during exercise. Circ Res 62:436–442

Shimokawa H, Flavahan NA, Vanhoutte PM (1989a) Natural course of the impairment of endothelium-dependent relaxation after balloon endothelium removal in porcine coronary arteries: possible dysfuntion of a pertussis-sensitive G protein. Circ Res 65:740–753

Shimokawa H, Flavahan NA, Sheperd JT, Vanhoutte PM (1989b) Endothelium-dependent inhibition of ergonovine-induced contraction is impaired in porcine coronary arteries with regenerated endothelium. Circulation 80:643–650

Siegel G, Schneider W (1981) Anions, cations, membrane potential and relaxation. In: Vanhoutte PM, Leusen I (eds) Vasodilatation. Raven, New York, pp 285–305

Silberbauer K, Slany J, Sinzinger H, Punzengruber C (1982) Molsidomin, Koronartherapeutikum mit plättchenaggregationshemmender Wirkung. Z Kardiol 71:539–543

Simmet T, Peskar BA (1986) Eicosanoids and the coronary circulation. Rev Physiol Biochem Pharmacol 104:1–64

Simon BC, Cunningham LD, Cohen RA (1990) Oxidized low density lipoproteins cause contraction and inhibit endothelium-dependent relaxation in pig coronary artery. J Clin Invest 86:75–79

Sink JC, Hill RC, Chitwood Jr WR, Abriss R, Wechsler AS (1979) Effects of phenylephrine on transmural distribution of myocardial blood flow in regions supplied by normal and collateral arteries during cardiopulmonary bypass. J Thorac Cardiovasc Surg 78:236–243

Smiesko V, Kozik J, Dolezel S (1985) Role of endothelium in the control of arterial diameter by blood flow. Blood Vessels 22:247–251

Smith JB, Araki H, Lefer AM (1980) Thromboxane A2, prostacyclin and aspirin: effects on vascular tone and platelet aggregation. Circulation 62 (Suppl V):V19-V25

Smitherman TC, Popma JJ, Said SI, Krejs GJ, Dehmer GJ (1989) Coronary hemodynamic effects of intravenous vasoactive intestinal peptide in humans. Am J Physiol 257:H1254–H1262

Spaan JAE, Breuls NPW, Laird JD (1981) Diastolic-systolic coronary flow differences are caused by intramyocardial pump action in anesthetized dog. Circ Res 49:584–593

Starke K (1981) Alpha-adrenoceptor subclassification. Rev Physiol Biochem Pharmacol 88:199–235

Steering Committee of the Physicians' Health Study Research Group (1989) Final report on the aspirin component of the ongoing physicians' health study. N Engl J Med 321:129–135

Steinberg D, Parthasarathy S, Carew TE, Khoo JC, Witztum JL (1989) Beyond cholesterol: modifications of low-density lipoprotein that increase its atherogenicity. N Engl J Med 320:915–924

Steinbrecher UP, Parthasarathy S, Leake DS, Witztum JL, Steinberg D (1984) Modification of low density lipoprotein by endothelial cells involves lipid peroxidation and degradation of low density lipoprotein phospholipids. Proc Natl Acad Sci USA 81:3883–3887

Stephenson JA, Gibson RE, Summers RJ (1988) An autoradiographic study of muscarinic cholinoceptors in blood vessels: no localization on vascular endothelium. Eur J Pharmacol 153:271–283

Stewart DJ, Holtz J, Pohl U, Bassenge E (1987a) Balance between endothelium-mediated dilating and direct constricting actions of serotonin on resistance vessels in the isolated rabbit heart. Eur J Pharmacol 143:131–134

Stewart DJ, Münzel T, Bassenge E (1987b) Reversal of acetylcholine-induced coronary resistance vessel dilation by hemoglobin. Eur J Pharmacol 136:239–242

Strader JR, Gwirtz PA, Jones CE (1988) Comparative effects of α_1- and α_2-adrenoceptors in modulation of coronary flow during exercise. J Pharmacol Exp Ther 246:772–778

Sugiura M, Inagami T, Kon V (1989) Endotoxin stimulates endothelin-release in vivo and in vitro as determined by radioimmunoassay. Biochem Biophys Res Comm 161: 1220–1227

Tabuchi Y, Nakamaru M, Rakugi H, Nagano M, Mikami H, Ogihara T (1989) Endothelin inhibits presynaptic adrenergic neurotransmission in rat mesenteric artery. Biochem Biophys Res Comm 161:803–808

Tesfamariam B, Cohen RA (1988) Inhibition of adrenergic vasoconstriction by endothelial cell shear stress. Circ Res 63:720–725

Tesfamariam B, Weisbrod RM, Cohen RA (1987) Endothelium inhibits responses of rabbit carotid artery to adrenergic nerve stimulation. Am J Physiol 253:H792–H798

Thaulow E, Guth BD, Schulz R, Ross Jr J (1989) Selective thromboxane A2 receptor blockade in experimental exercise-induced myocardial ischemia in dogs. Acta Physiol Scand 136:321–330

Thiemermann C, Smith III EF, Schrör K (1986) Successful treatment of acute myocardial ischemia with teopranitol – a novel organic nitrate. Eur Heart J 7:418–424

Tillmanns H, Ikeda S, Hansen H, Sarma JSM, Fauvel J-M, Bing RJ (1974) Microcirculation in the ventricle of the dog and turtle. Circ Res 34:561–569

Toda N (1986) Alpha-adrenoceptor subtypes and diltiazem actions in isolated human coronary arteries. Am J Physiol 250:H718–H724

Tomoike H, Egashira K, Yamamoto Y, Nakamura M (1989) Enhanced responsiveness of smooth muscle, impaired endothelium-dependent relaxation and the genesis of coronary spasm. Am J Cardiol 63:33E–39E

Toyo-oka T, Aizawa T, Suzuki N, Hirata Y, Miyauchi T, Yanagisawa M, Masaki T (1989) Contribution of endothelin (ET) and atrial natriuretic factor (ANF) to coronary spasm in patients with vasospastic angina. Circulation 80 (Suppl II):II–126 (abstract)

Tzivoni D, Keren A, Benhorin J, Gottlieb S, Atlas D, Stern S (1983) Prazosin therapy for refractory variant angina. Am Heart J 105:262–266

Uchida Y, Murao S (1974) Cyclic changes in peripheral blood pressure of partially constricted coronary artery. Jpn Coll Angiol 14:383–393

Van Gilst WH, van Wijngaarden J, Scholtens E, de Graeff PA, de Langen CDJ, Wesseling H (1987) Captopril-induced increase in coronary flow: an SH-dependent effect on arachidonic acid metabolism? J Cardiovasc Pharmacol 9 (Suppl 2):S31–S36

Van Gilst WH, Tio RA, de Graeff PA, van Wijngaarden J, Scholtens E, de Langen CDJ, Wesseling H (1989) Antiischämische Wirkungen von Conversions-Enzym-Hemmern. Muench Med Wschr 131 (Suppl 1):S27–S30

Vanhoutte PM (1987) Endothelium and the control of vascular tissue. News Pharmacol Sci 2:18–22

Vanhoutte PM (1988) The endothelium and control of coronary arterial tone. Hosp Pract 77–94

Vanhoutte PM, Houston DS (1985) Platelets, endothelium, and vasospasm. Circulation 72:728–734

Vanhoutte PM, Katusic ZS, Shepherd JT (1984) Vasopressin induces endothelium-dependent relaxations of cerebral and coronary, but not of systemic arteries. J Hypertension 2 (Suppl 3):421–422

Van Meel JCA, de Jonge A, Timmermans PBMWM, van Zwieten PA (1981) Selectivity of some alpha adrenoceptor agonists for peripherial alpha$_1$- and alpha$_2$-adrenoceptors in the normotensive rat. J Pharmacol Exp Ther 219:760–767

Van Winkle DM, Feigl EO (1989) Acetylcholine causes coronary vasodilation in dogs and baboons. Circ Res 65:1580–1593

Van Zwieten PA, Timmermans PBMWM (1983) Cardiovascular α_2-receptors. J Mol Cell Cardiol 15:717–733

Vatner DE, Knight DR, Homcy CJ, Vatner SF, Young MA (1986) Subtypes of β-adrenergic receptors in bovine coronary arteries. Circ Res 59:463–473

Vatner SF (1980) Correlation between acute reductions in myocardial blood flow and function in conscious dogs. Circ Res 47:201–207

Vatner SF, Hintze TH (1983) Mechanism of constriction of large coronary arteries by β-adrenergic receptor blockade. Circ Res 53:389–400

Vatner SF, McRitchie RJ (1975) Interaction of the chemoreflex and the pulmonary inflation reflex in the regulation of coronary circulation in conscious dogs. Circ Res 37:664–673

Vatner SF, Hintze TH, Macho P (1982) Regulation of large coronary arteries by β-adrenergic mechanisms in the conscious dog. Circ Res 51:56–66

Vekshtein VI, Yeung AC, Vita JA, Nabel EG, Fish RD, Bittl JA, Selwyn AP, Ganz P (1989) Fish oil improves endothelium-dependent relaxation in patients with coronary artery disease. Circulation 80 (Suppl II):II–434 (abstract)

Verbeuren TJ, Jordaens FH, Zonnekeyn LL, Van Hove CE, Coene MC, Herman AG (1986) Effect of hypercholesterolemia on vascular reactivity in the rabbit I. Endothelium-dependent and endothelium-independent contractions and relaxations in isolated arteries of control and hypercholesterolemic rabbits. Circ Res 58:552–564

Verbeuren TJ, Jordaens FH, Van Hove CE, von Hoydonck AE, Herman AG (1990) Release and vascular activity of the endothelium-derived relaxing factor in atherosclerotic aorta. Eur J Pharmacol (in press)

Vita JA, Treasure CB, Nabel EG, McLenachan JM, Fish RD, Yeung AC, Vekshtein VI, Selwyn AP, Ganz P (1990) Coronary vasomotor response to acetylcholine relates to risk factors for coronary artery disease. Circulation 81:491–497

Vlahakes GJ, Baer RW, Uhlig PN, Verrier ED, Bristow JD, Hoffman JIE (1982) Adrenergic influence in the coronary circulation of conscious dogs during maximal vasodilation with adenosine. Circ Res 51:371–384

Von Restorff W, Bassenge E (1977) Transient effects of norepinephrine on myocardial oxygen balance. Pfluegers Arch 370:131–137

Von Restorff W, Holtz J, Bassenge E (1977) Exercise induced augmentation of myocardial oxygen extraction in spite of normal coronary dilatory capacity in dogs. Pfluegers Arch 372:181–185

Vrints C, Verbeuren TJ, Snoeck J, Herman AG (1990) Effects of hypercholesterolemia on coronary vascular reactivity. In: Rubanyi GM, Vanhoutte PM (eds) Endothelium-derived contracting factors. Karger, Basel, pp 162–168

Wainwright CL, Parratt JR (1988) The effect of L655,240, a selective thromboxane and prostaglandin endoperoxide antagonist, on ischemia- and reperfusion-induced cardiac arrhythmias. J Cardiovasc Pharmacol 12:264–271

Wang H-H, Katz RL (1965) Effects of changes in coronary blood pH on the heart. Circ Res 17:114–122

Wargovich T, Mehta J, Nichols WW, Pepine CJ, Conti CR (1985) Reduction in blood flow in normal and narrowed coronary arteries of dogs by leukotriene C4. J Am Coll Cardiol 6:1047–1051

Weitzell R, Tanaka T, Starke K (1979) Pre- and postsynaptic effects of yohimbine stereoisomers on noradrenergic transmission in the pulmonary artery of the rabbit. Naunyn Schmiedebergs Arch Pharmacol 308:127–136

Wennmalm A, Lanne B, Petersson A-S (1990) Detction of endothelium-derived relaxing factor in human plasma in the basal state and following ischemia using electron paramagnetic resonance spectrometry. Anal Biochem 197: (in press)

Werns SW, Walton JA, Hsia HH, Nabel EG, Sanz ML, Pitt B (1989) Evidence of endothelial dysfunction in angiographically normal coronary arteries of patients with coronary artery disease. Circulation 79:287–291

Wiggers CJ (1954) The interplay of coronary vascular resistance and myocardial compression in regulating coronary flow. Circ Res 2:271–279

Wilson FR, Marcus ML, White CW (1988) Pulmonary inflation reflex: its lack of physiological significance in coronary circulation of humans. Am J Physiol 255:H866−H871

Wilson FR, Lesser JR, Laxson DD, White CW (1989) Intense microvascular constriction after angioplasty of acute thrombotic coronary arterial lesions. Lancet I:807−811

Winniford MD, Filipchuk N, Hillis LD (1983) Alpha-adrenergic blockade for variant angina: a long-term, double-blind, randomized trial. Circulation 67:1185−1188

Woodman OL, Vatner SF (1987) Coronary vasoconstriction mediated by α_1- and α_2-adrenoceptors in conscious dogs. Am J Physiol 253:H388−H393

Wright CD, Mülsch A, Busse R, Osswald H (1989) Generation of nitric oxide by human neutrophils. Biochem Biophys Res Comm 160:813−819

Wright CE, Angus JA (1986) Effects of hypertension and hypercholesterolemia on vasodilatation in the rabbit. Hypertension 8:361−371

Wüsten B, Buss DD, Deist H, Schaper W (1977) Dilatory capacity of the coronary circulation and its correlation to the arterial vasculature in the canine left ventricle. Basic Res Cardiol 72:636−650

Yamaguchi N, DeChamplain J, Nadeau RA (1977) Regulation of norephinephrine release from cardiac sympathetic fibers in the dog by presynaptic α- and β-receptors. Circ Res 41:108−117

Yamamoto Y, Tomoike H, Egashira K, Nakamura M (1987a) Attenuation of endothelium-related relaxation and enhanced responsiveness of vascular smooth muscle to histamine in spastic coronary arterial segments from miniature pigs. Circ Res 61:772−778

Yamamoto Y, Tomoike H, Egashira K, Kobayashi T, Kawasaki T, Nakamura M (1987b) Pathogenesis of coronary artery spasm in miniature swine with regional intimal thickening after balloon denudation. Circ Res 60:113−121

Yanagisawa M, Masaki T (1989a) Molecular biology and biochemistry of the endothelins. Trends Pharmacol Sci 10:374−378

Yanagisawa M, Masaki T (1989b) Endothelin, a novel endothelium-derived peptide. Pharmacological activities, regulation and possible roles in cardiovascular control. Biochem Pharmacol 38:1877−1883

Yanagisawa M, Kurihara H, Kimura S, Goto K, Masaki T (1988) Endothelium-derived novel vasoconstrictor peptide endothelin: a possible endogenous agonist for voltage-dependent Ca^{2+} channels. In: Morad M, Nayler W, Kazda S, Schramm M (eds) The calcium channel: structure, function and implications. Springer, Berlin Heidelberg New York, pp 575−585

Yasue H, Touyama M, Kato H, Tanaka S, Akiyama F (1976) Prinzmetal's variant form of angina as a manifestation of alpha-adrenergic receptor-mediated coronary artery spasm: documentation by coronary arteriography. Am Heart J 91:148−155

Ylä-Herttuala S, Palinski W, Rosenfeld ME, Parthasarathy S, Carew TE, Butler S, Witztum JL, Steinberg D (1989) Evidence for the presence of oxidatively modified low density lipoproteins in atherosclerotic lesions of rabbit and man. J Clin Invest 84:1086−1095

Yoshizumi M, Kurihara H, Sugiyama T, Takaku F, Yanagisawa M, Masaki T, Yazaki Y (1989) Hemodynamic shear stress stimulates endothelin production by cultured endothelial cells. Biochem Biophys Res Comm 161:859−864

Young MA, Knight DR, Vatner SF (1987) Autonomic control of large coronary arteries and resistance vessels. Prog Cardiovasc Dis 30:211−234

Young MA, Knight DR, Vatner SF (1988a) Parasympathetic coronary vasoconstriction induced by nicotine in conscious calves. Circ Res 62:891−895

Young MA, Vatner DE, Knight DR, Graham RM, Homcy CJ, Vatner SF (1988b) α-adrenergic vasoconstriction and receptor subtypes in large coronary arteries of calves. Am J Physiol 255:H1452−H1459

Zeiher AM, Drexler H, Wollschläger H, Just H (1989a) Preserved flow-mediated vasodilation despite acetylcholine-induced vasoconstriction in atherosclerotic coronary arteries in man. J Am Coll Cardiol 13 (Suppl A):132A (abstract)

Zeiher AM, Drexler H, Wollschlaeger H, Saurbier B, Just H (1989b) Coronary vasomotion in response to sympathetic stimulation in humans: importance of the functional integrity of the endothelium. J Am Coll Cardiol 14:1181–1190

Zimmerman BG (1978) Actions of angiotensin on adrenergic nerve endings. Fed Proc 37:199–202

Zucker IH, Cornish KG, Hackley J, Bliss K (1987) Effects of left ventricular receptor stimulation on coronary blood flow in conscious dogs. Circ Res 61 (Suppl II):II54–II60

Subject Index